ARTIFICIAL NEURAL NETWORKS IN BIOLOGICAL AND ENVIRONMENTAL ANALYSIS

ANALYTICAL CHEMISTRY SERIES

Series Editor

Charles H. Lochmüller

Duke University

ANALYTICAL CHEMISTRY SERIES

ARTIFICIAL NEURAL NETWORKS IN BIOLOGICAL AND ENVIRONMENTAL ANALYSIS

Grady Hanrahan

CRC Press
Taylor & Francis Group
Boca Raton   London   New York

CRC Press is an imprint of the
Taylor & Francis Group, an **informa** business

CRC Press
Taylor & Francis Group
6000 Broken Sound Parkway NW, Suite 300
Boca Raton, FL 33487-2742

First issued in paperback 2017

ISBN 13: 978-1-138-11293-3 (pbk)
ISBN 13: 978-1-4398-1258-7 (hbk)

Library of Congress Cataloging-in-Publication Data

Hanrahan, Grady.
    Artificial neural networks in biological and environmental analysis / Grady Hanrahan. -- 1st ed.
    p. cm.
    Summary: "Drawing on the experience and knowledge of a practicing professional, this book provides a comprehensive introduction and practical guide to the development, optimization, and application of artificial neural networks (ANNs) in modern environmental and biological analysis. Based on our knowledge of the functioning human brain, ANNs serve as a modern paradigm for computing. Presenting basic principles of ANNs together with simulated biological and environmental data sets and real applications in the field, this volume helps scientists comprehend the power of the ANN model to explain physical concepts and demonstrate complex natural processes"-- Provided by publisher.
    Summary: "The cornerstones of research into prospective tools of artificial intelligence originate from knowledge of the functioning brain. Like most transforming scientific endeavors, this field-- once viewed with speculation and doubt--has had profound impacts in helping investigators elucidate complex biological, chemical, and environmental processes. Such efforts have been catalyzed by the upsurge in computational power and availability, with the co-evolution of software, algorithms, and methodologies contributing significantly to this momentum. Whether or not the computational power of such techniques is sufficient for the design and construction of truly intelligent neural systems is of continued debate. In writing Artificial Neural Networks in Biological and Environmental Analysis, my aim was to provide in-depth and timely perspectives on the fundamental, technological, and applied aspects of computational neural networks. By presenting basic principles of neural networks together with real applications in the field, I seek to stimulate communication and partnership among scientists in the fields as diverse as biology, chemistry, mathematics, medicine, and environmental science. This interdisciplinary discourse is essential not only for the success of independent and collaborative research and teaching programs, but also for the continued acquiescence of the use of neural network tools in scientific inquiry"-- Provided by publisher.
    Includes bibliographical references and index.
    ISBN 978-1-4398-1258-7 (hardback)
    1. Artificial intelligence--Biological applications. 2. Biology--Data processing. 3. Environmental engineering--Data processing. 4. Neural networks (Computer science)--Scientific applications. I. Title.

QH324.25.H36 2011
570.285'63--dc22
                                                  2010038461

Visit the Taylor & Francis Web site at
http://www.taylorandfrancis.com

and the CRC Press Web site at
http://www.crcpress.com

To my dearest mother
In memory of Dr. Ira Goldberg

Contents

Foreword

The sudden rise in popularity of artificial neural networks (ANNs) during the 1980s and 1990s indicates that these techniques are efficient in solving complex chemical and biological problems. This is due to characteristics such as robustness, fault tolerance, adaptive learning, and massively parallel analysis capabilities. ANNs have been featured in a wide range of scientific journals, often with promising results.

It is frequently asked whether or not biological and environmental scientists need more powerful statistical methods than the more traditional ones currently employed in practice. The answer is yes. Scientists deal with very complicated systems, much of the inner workings of which are, as a rule, unknown to researchers. If we only use simple, linear mathematical methods, information that is needed to truly understand natural systems may be lost. More powerful models are thus needed to complement modern investigations. For example, complex biological problems such as alignment and comparison of DNA and RNA, gene finding and promoter identification from DNA sequencing, enzymatic activities, protein structure predictions and classifications, etc., exist that fall within the scope of bioinformatics. However, the development of new algorithms to model such processes is needed, in which ANNs can play a major role. Moreover, human beings are concerned about the environment in which they live and, therefore, numerous research groups are now focusing on developing robust methods for environmental analysis.

It is not an easy task to write a book that presents a corpus of existing knowledge in a discipline and yet also keeps a close enough watch on the advancing front of scientific research. The task is particularly difficult when the range of factual knowledge is vast, as it is in the environmental and biological sciences. As a consequence, it is difficult to deal adequately with all the developments that have taken place during the past few decades within a single text. Dr. Grady Hanrahan has nevertheless managed to review the most important developments to some degree and achieve a satisfactory overall balance of information. Students and biological and environmental scientists wishing to pursue the neural network discipline will find a comprehensive introduction, along with indications where more specialized accounts can be found, expressed in clear and concise language, with some attention given to current research interests. A number of artificial neural network texts have appeared in recent years, but few, if any, present as harmonious a balance of basic principles and diverse applications as does this text, for which I feel privileged to write this foreword.

Mehdi Jalali-Heravi
Chemometrics and Chemoinformatics Research Group
Sharif University of Technology

Preface

The cornerstones of research into prospective tools of artificial intelligence originate from knowledge of the functioning brain. Similar to most transforming scientific endeavors, this field—once viewed with speculation and doubt—has had a profound impact in helping investigators elucidate complex biological, chemical, and environmental processes. Such efforts have been catalyzed by the upsurge in computational power and availability, with the co-evolution of software, algorithms, and methodologies contributing significantly to this momentum. Whether or not the computational power of such techniques is sufficient for the design and construction of truly intelligent neural systems is the subject of continued debate. In writing *Artificial Neural Networks in Biological and Environmental Analysis*, my aim was to provide in-depth and timely perspectives on the fundamental, technological, and applied aspects of computational neural networks. By presenting the basic principles of neural networks together with real-world applications in the field, I seek to stimulate communication and partnership among scientists in fields as diverse as biology, chemistry, mathematics, medicine, and environmental science. This interdisciplinary discourse is essential not only for the success of independent and collaborative research and teaching programs, but also for the continued interest in the use of neural network tools in scientific inquiry.

In the opening chapter, an introduction and brief history of computational neural network models in relation to brain functioning is provided, with particular attention being paid to individual neurons, nodal connections, and transfer function characteristics. Building on this, Chapter 2 details the operation of a neural network, including discussions of neuron connectivity and layer arrangement. Chapter 3 covers the eight-step development process and presents the basic building blocks of model design, selection, and application from a statistical perspective. Chapter 4 was written to provide readers with information on hybrid neural approaches including neuro-fuzzy systems, neuro-genetic systems, and neuro-fuzzy-genetic systems, which are employed to help achieve increased model efficiency, prediction, and accuracy in routine practice. Chapters 5 and 6 provide a glimpse of how neural networks function in real-world applications and how powerful they can be in studying complex natural processes. Included in Chapter 6 is a subsection contribution by Błażej Kudłak and colleagues titled "Neural Networks and the Evolution of Environmental Change." The basic fundamentals of matrix operations are provided in Appendix I. In addition, working data sets of selected applications presented in Chapters 5 and 6 are supplied in Appendices II and III, respectively.

This book is by no means comprehensive, but it does cover a wealth of important theoretical and practical issues of importance to those incorporating, or wishing to incorporate, neural networks into their academic, regulatory, and industrial pursuits. In-depth discussion of mathematical concepts is avoided as much as possible, but appropriate attention has been given to those directly related to neuron function, learning, and statistical analysis. To conclude, it is my hope that you will find this book interesting and enjoyable.

Grady Hanrahan
Los Angeles, California

Acknowledgments

There are numerous people I must acknowledge for my success in completing this daunting task. First and foremost I thank my family for their continued support of my often consuming academic endeavors. I am forever indebted to Professor Paul Worsfold, my Ph.D. dissertation advisor, who gave me the freedom to write during my studies and has continued to support my activities over the years. I am grateful to Senior Editor Barbara Glunn (CRC Press) for believing in the concept of this book when I approached her with my idea. I also thank Project Coordinator David Fausel and the CRC Press editorial support staff for a superlative job of editing and formatting, and for seeing this book on through to final production in a timely and professional manner. A number of organizations have granted permission to reprint or adapt materials originally printed elsewhere, including the American Chemical Society, Elsevier, John Wiley & Sons, Oxford University Press, and Wiley-VCH. I thank Błażej Kudłak and his colleagues for the addition of valuable information in Chapter 6.

I thank the countless number of students with whom I have worked on various neural network applications, including Jennifer Arceo, Sarah Muliadi, Michael Jansen, Toni Riveros, Jacqueline Kiwata, and Stephen Kauffman. I am grateful to Jennifer Arceo and Vicki Wright for their help with literature searches and formatting of book content. I thank Kanjana Patcharaprasertsook for the illustrations contained in this book. Finally, I thank my collaborators Drs. Frank Gomez, Krishna Foster, Mehdi Jalali-Heravi, and Edith Porter for their continued interest in this field.

The Author

Grady Hanrahan received his Ph.D. in environmental analytical chemistry from the University of Plymouth, U.K. With experience in directing undergraduate and graduate research, he has taught analytical chemistry and environmental science at California State University, Los Angeles, and California Lutheran University. He has written or co-written numerous peer-reviewed technical papers and is the author and editor of five books detailing the use of modern chemometric and modeling techniques to solve complex biological and environmental problems.

Guest Contributors

The following individuals contributed material to Chapter 6 (Section 6.2.5):

Błażej Kudłak
Department of Analytical Chemistry
Faculty of Chemistry
Gdańsk University of Technology
Gdańsk, Poland

Robert Kudłak
Institute of Socio-Economic Geography
and Spatial Management
Faculty of Geographical and Geological
Sciences
Adam Mickiewicz University
Poznań, Poland

Jacek Namieśnik
Department of Analytical Chemistry
Faculty of Chemistry
Gdańsk University of Technology
Gdańsk, Poland

Vasil Simeonov
Department of Analytical Chemistry
Faculty of Chemistry
University of Sofia "St. Kl. Okhridski"
Sofia, Bulgaria

Stefan Tsakovski
Department of Analytical Chemistry
Faculty of Chemistry
University of Sofia "St. Kl. Okhridski"
Sofia, Bulgaria

Glossary of Acronyms

AI	Artificial intelligence
AIC	Akaike information criterion
ANFIS	Adaptive neuro-fuzzy inference systems
ANN	Artificial neural network
ANOVA	Analysis of variance
BIC	Bayesian information criterion
BP-BM	Back-propagation algorithm with back update
CE	Capillary electrophoresis
EMMA	Electrophoretically mediated microanalysis
ES	Evolving strategies
FTIR	Fourier transform infrared spectroscopy
GA	Genetic algorithm
GLM	Generalized linear models
KBS	Knowledge-based systems
k-NN	k nearest-neighbor method
LM	Levenberg–Marquardt algorithm
LMS	Least mean square
LVQ	Learning vector quantization
MAE	Mean absolute error
MAR	Missing at random
MCAR	Missing completely at random
MCMC	Markov chain Monte Carlo method
MLR	Multiple linear regression
MNAR	Missing not at random
MRE	Mean relative error
MSE	Mean square error
MSPD	Matrix solid-phase dispersion
NIC	Network information criterion
NRBRNN	Normalized radial basis neural networks
OCW	Overall connection weights
OGL	Ordinary gradient learning
PC	Principal component
PCA	Principal component analysis
PCR	Principal component regression
PDF	Probability density function
PLS	Partial least squares
PNN	Probabilistic neural networks
QSAR	Quantitative structure–activity relationship

RBF	Radial basis functions
RBFN	Radial basis function networks
RC	Relative contribution
RMSE	Root mean square error
RMSEF	Root mean square error for fitting
RMSEP	Root mean square error for prediction
RNN	Recurrent neural networks
SANGL	Adaptive natural gradient learning with squared error
SOM	Self-organizing maps
SSE	Sum of squared error
SVM	Support vector machine
UBF	Universal basis functions
WT	Wavelet transform

1 Introduction

Because evolution has no such plan, it becomes relevant to ask whether the ability of large collections of neurons to perform "computational" tasks may in part be a spontaneous collective consequence of having a large number of interacting simple neurons.

J.J. Hopfield
Proceedings of the National Academy of Sciences USA, 1982

1.1 ARTIFICIAL INTELLIGENCE: COMPETING APPROACHES OR HYBRID INTELLIGENT SYSTEMS?

Minsky and Papert (1969) in their progressive and well-developed writing discussed the need to construct artificial intelligence (AI) systems from diverse components: a requisite blend of symbolic and connectionist approaches. In the symbolic approach, operations are performed on symbols, where the physical counterparts of the symbols, and their structural properties, dictate a given system's behavior (Smolensky, 1987; Spector, 2006). It is argued that traditional symbolic AI systems are rigid and specialized, although there has been contemporary development of symbolic "learning" systems employing fuzzy, approximate, or heuristic components of knowledge (Xing et al., 2003) to counteract this narrow view.

The connectionist approach is inspired by the brain's neural structure and is generally regarded as a learning systems approach. Connectionist systems are characterized as having parallel processing units that exhibit intelligent behavior without structured symbolic expressions (Rumelhart and McClelland, 1986; Spector, 2006). Learning proceeds as a result of the adjustment of weights within the system as it performs an assigned task. Critics of this approach do question whether the computational power of connectionist systems is sufficient for the design and construction of truly intelligent systems (Smolensky, 1987; Chalmers, 1996). On a more basic level, the question is posed whether or not they can in fact compute. Piccinini (2004, 2008) endeavored to address this issue in a well-reasoned paper detailing connectionist systems. More exclusively, two distinctions were drawn and applied in reference to their ability to compute: (1) those between classical and nonclassical computational concepts and (2) those between connectionist computation and other connectionist processes. He argued that many connectionist systems do in fact compute through the manipulation of strings of digits in harmony with a rule delineated over the inputs. Alternatively, specific connectionist systems (e.g., McCulloch–Pitts nets [defined shortly]) compute in a more classical way by operating in accordance with a given algorithm for generating successive strings of digits. Furthermore, he argues that other connectionist systems compute in a trainable, nonclassical way by turning their inputs into their outputs by virtue of their continuous dynamics. There is thus a continued debate as to which system—classical or nonclassical, computational or noncomputational—best mimics the brain. Piccinini pointed to those connectionist theorists who agree with classicists that brains perform computations, and neural computations explain cognition in some form or fashion (e.g., Hopfield, 1982; Rumelhart and McClelland, 1986; Churchland, 1989; Koch, 1999; Shagrir, 2006). He gave equal coverage to those classicists who argue that nonclassical connectionist systems do not perform computations at all (e.g., Fodor, 1975; Gallistel and Gibbon, 2002), and a separate group of connectionist theorists who deny the fact that brains are capable of even limited computation (e.g., Edelman, 1992; Freeman, 2001).

It is then appropriate to ask: Are symbolic and connectionist approaches functionally different and contradictory in nature? Paradoxically, do they appropriately coalesce to complement each other's strengths to facilitate emulation of human cognition through information processing, knowledge representation, and directed learning? There is great support and movement toward hybrid systems: the combination of two or more techniques (paradigms) to realize convincing problem-solving strategies. The suitability of individual techniques is case specific, with each having distinct advantages and potential drawbacks. Characteristically, hybrid systems will combine two or more techniques with the decisive objective of gaining strengths and overcoming the weaknesses of single approaches.

Three prevalent types of hybrid systems are reported (Chen et al., 2008):

1. Sequential—a process by which the first paradigm passes its output to a second of subsequent output generation
2. Auxiliary—a process by which the first paradigm obtains given information from a second to generate an output
3. Embedded—a process by which two paradigms are contained within one another

Consider the integration of symbolic (e.g., fuzzy systems) and connectionist (e.g., neural networks) systems. This embedded combination, toward a neuro-fuzzy system, provides an effective and efficient approach to problem solving. Fuzzy systems carry a notion that truth values (in fuzzy logic terms) or membership values (in fuzzy sets) are indicated as a range [0.0, 1.0], with 0.0 representing absolute falseness and 1.0 representing absolute truth (Dubois and Prade, 2004). Fuzzy systems make use of linguistic knowledge and are interpretable in nature. In contrast, neural networks are largely considered a "black box" approach and characteristically learn from scratch (Olden and Jackson, 2002). By combining these two paradigms, the drawbacks pertaining to both become complementary. A variety of other hybrid approaches are used, including expanded hybrid connectionist-symbolic models, evolutionary neural networks, genetic fuzzy systems, rough fuzzy hybridization, and reinforcement learning with fuzzy, neural, or evolutionary methods, and symbolic reasoning methods. A variety of these models will be discussed in Chapter 4 and in various applications presented throughout this book.

1.2 NEURAL NETWORKS: AN INTRODUCTION AND BRIEF HISTORY

Neural network foundational concepts can be traced back to seminal work by McCulloch and Pitts (1943) on the development of a sequential logic model of a neuron. Although the principal subject of this paper was the nervous system and neuron function, the authors presented simplified diagrams representing the functional relationships between neurons conceived as binary elements. Their motivation, spurred by philosophy, logic, and mathematics, led to the development of formal assumptions, theoretical presuppositions, and idealizations based on general knowledge of the nervous system and nerve cells. Presumably, their goal was to develop formal logistic physiology operations in the brain, but at the level of the neuron. A more detailed look into McCulloch–Pitts nets (as they are commonly termed) can be found in an informative review by Cowan (1990). In it, Cowan describes how such nets "embody the logic of propositions and permit the framing of sharp hypotheses about the nature of brain mechanisms, in a form equivalent to computer programs." He provides a formalized look at McCulloch–Pitts nets and their logical representation of neural properties, detailed schematics of logic functions, and valuable commentary and historical remarks on McCulloch–Pitts and related neural networks from 1943 to 1989.

In the late 1950s, Rosenblatt and others were credited with the development of a network based on the perceptron: a unit that produces an output scaled as 1 or −1 depending on the weighted combination of inputs (Marinia et al., 2008). Rosenblatt demonstrated that McCulloch–Pitts networks with modifiable connections could be "trained" to classify certain sets of patterns as similar or distinct (Cowan, 1990). Perceptron-based neural networks were considered further by Widrow and Hoff (1960). Their version, termed Adaline (for adaptive linear neuron), was a closely related version of the perceptron, but differed in its approach to training. Adalines have been reported to match closely the performance of perceptrons in a variety of tasks (Cowan, 1990). In 1969, Minsky and Papert discussed the inherent limitations of perceptron-based neural networks in their landmark book, *Perceptrons*. Discussions of these limitations and efforts to remedy such concerns will be covered in subsequent chapters of this book. Of particular significance is the work by Hopfield (1982), who introduced statistical mechanics to explain the operation of a class of recurrent networks that could ultimately be used as an associative memory (Hagan et al., 1996). Hopfield summarized:

> Memories are retained as stable entities or Gestalts and can be correctly recalled from any reasonably sized subpart. Ambiguities are resolved on a statistical basis. Some capacity for generalization is present, and time ordering of memories can also be encoded. These properties follow from the nature of the flow in phase space produced by the processing algorithm, which does not appear to be strongly dependent on precise details of the modeling. This robustness suggests that similar effects will obtain even when more neurobiological details are added. (Hopfield, 1982)

In 1985, Hopfield and Tank proposed a neural network approach for use in optimization problems, which attracted many new users to the neural computing field. There are countless others who have made significant contributions in this field, including Amari, Cooper, Fukushima, Anderson, and Grossberg, to name a few. Since these major milestones, neural networks have experienced an explosion of interest (but not without criticism) and use across disciplines, and are arguably the most widely used connectionist approaches employed today.

When we think of a neural network model, we are referring to the network's arrangement; related are neural network algorithms: computations that ultimately produce the network outputs (Jalali-Heravi, 2008). They are a modern paradigm based on computational units that resemble basic information processing properties of biological neurons, although in a more abstract and simplified manner. The key feature of this paradigm is the structure of the novel information processing system: a working environment composed of a large number of highly interconnected processing elements (neurons) working in unison to solve user-specific problems. They can be used to gain information regarding complex chemical and physical processes; predict future trends; collect, interpret, and represent data; and solve multifaceted problems without necessarily creating a model of a real biological system.

Neural networks have the property of *learning* by example, similar to and patterned after biological systems and the adjustments to the synaptic connections that exist between individual neurons (Luger and Stubblefield, 1997). A second fundamental property of neural networks is their ability to implement nonlinear functions by allowing a uniform approximation of any continuous function. Such a property is fundamental in studying biological and environmental systems, which may exhibit variable responses even when the input is the same. Nevertheless, there are reported obstacles to the success of neural network models and their general applicability. Neural networks are statistical models that use nonparametric approaches. Thus, a priori knowledge is not obviously to be taken into account any more than a posterior knowledge (Oussar and Dreyfus, 2001; Johannet et al., 2007). Therefore, neural networks are often treated as the aforementioned black box representation (pictured schematically in Figure 1.1) whose inner workings are concealed from the researcher, thus making it challenging to authenticate how explicit decisions are acquired. It is generally considered that information stored in neural networks is a set of weights and connections that provide no insight into how a task is actually performed. Conversely, recent studies have shown that by using various techniques the black box can be opened, or at least provide gray box solutions. Techniques such as sensitivity

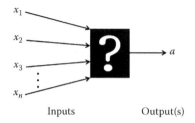

Inputs   Output(s)

FIGURE 1.1   Black box models employ user-defined inputs ($x_1, x_2, x_3, ..., x_n$) to return output (a) to the user. Note that there can be more than one output depending on the application. The "black box" portion of the system contains formulas and calculations that the user does not see or necessarily need to know to use the system.

analysis (Recknagel et al., 1997; Scardi, 2001), input variable relevances and neural interpretation diagrams (Özesmi et al., 2006), random tests of significance (Olden and Jackson, 2002), fuzzy set theory (Peters et al., 2009), and partial derivatives (Rejyol et al., 2001) have been used to advance model transparency. Detailed information on how current research focuses on implementing alternative approaches such as these inside the network will be detailed later in this book.

1.2.1 THE BIOLOGICAL MODEL

The biological model provides a critical foundation for creating a functional mathematical model. An understanding of neuronal and synaptic physiology is important, with nervous system complexity being dependent on the interconnections between neurons (Parker and Newsom, 1998). Four main regions comprise a prototypical neuron's structure (Figure 1.2): the soma (cell body), dendrites, axons, and synaptic knobs. The soma and dendrites represent the location of input reception, integration,

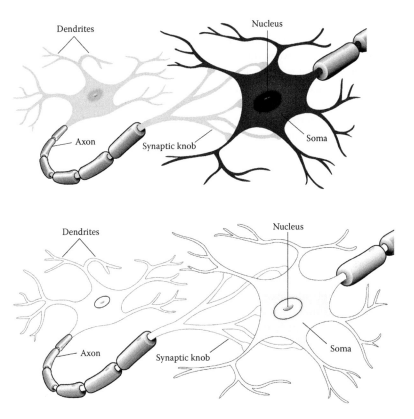

FIGURE 1.2 *A color version of this figure follows page 106.* Biological neurons organized in a connected network, both receiving and sending impulses. Four main regions comprise a neuron's structure: the soma (cell body), dendrites, axons, and synaptic knobs. (From Hanrahan, G. 2010. *Analytical Chemistry,* 82: 4307–4313. With permission from the American Chemical Society.)

and coordination of signals arising from presynaptic nerve terminals. Information (signal) propagation from the dendrite and soma occurs from the axon hillock and down its length. Such signals are termed action potentials, the frequency of action potential generation being proportional to the magnitude of the net synaptic response at the axon hillock (Giuliodori and Zuccolilli, 2004). Recent evidence has suggested the existence of bidirectional communication between astrocytes and neurons (Perea and Avaque, 2002). Astrocytes are polarized glial cells strongly coupled to one another by gap junctions that provide biochemical support for endothelial cells. Recent data also suggest that astrocytes signal to neurons via the Ca^{2+}-dependent release of glutamate (Bennett et al., 2003). As a consequence of this evidence, a new concept of synaptic physiology has been proposed.

There are three fundamental concepts that are important in understanding brain function and, ultimately, the construction of artificial neural networks. First, the strength of the connection between two neurons is vital to memory function; the connections will strengthen, wherein an increase in synaptic efficacy arises from the presynaptic cell's repeated stimulation of the postsynaptic cell (Paulsen and Sejnowski, 2000). This mechanism for synaptic plasticity describes Hebb's rule, which states that the simultaneous excitation of two neurons results in a strengthening of the connections between them (Hebb, 1949). Second, the amount of excitation (increase in the firing rates of connected neurons) or inhibition (decrease in the firing rates of connected neurons) is critical in assessing neural connectivity. Generally, a stronger connection results in increased inhibition or excitation. Lastly, the transfer function is used in determining a neuron's response. The transfer function describes the variation in neuron firing rate as it receives desired inputs. All three concepts must be taken into account when describing the functioning properties of neural networks.

1.2.2 THE ARTIFICIAL NEURON MODEL

A neural network is a computing paradigm patterned after the biological model discussed earlier. It consists of interconnected processing elements called nodes or neurons that work together to produce an output function. The output of a neural network relies on the functional cooperation of the individual neurons within the network, where processing of information is characteristically done in parallel rather than sequentially as in earlier binary computers (Hanrahan, 2010). Consider the multiple-input neuron presented in Figure 1.3 for a more detailed examination of artificial neuron function. The individual scalar inputs x_1, x_2, x_3, ..., x_n are each weighted with appropriate elements w_1, w_2, w_3, ...,w_n of the weight matrix **W**. The sum of the weighted inputs and the bias forms the net input n, proceeds into a transfer function f, and produces the scalar neuron output a written as (Hagan et al., 1996)

$$a = f(\mathbf{W}x + b) \tag{1.1}$$

If we again consider the biological neuron pictured earlier, the weight w corresponds to synapse strength, the summation represents the cell body and the transfer function, and a symbolizes the axon signal.

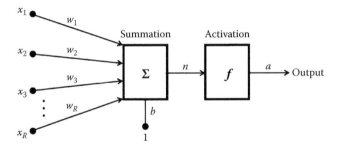

FIGURE 1.3 A basic multiple-input artificial neuron model. Individual scalar inputs are weighted appropriate elements w_1, w_2, w_3, ..., w_n of the weight matrix W. The sum of the weighted inputs and the bias forms the net input n, proceeds into a transfer function f, and produces the scalar neuron output a. (From Hanrahan, G. 2010. *Analytical Chemistry*, 82: 4307–4313. With permission from the American Chemical Society.)

A more concentrated assessment of the transfer function reveals greater insight into the way signals are processed by individual neurons. This function is defined in the N-dimensional input space, also termed the parameter space. It is composed of both an activation function (determines the total signal a neuron receives) and output function. Most network architectures start by computing the weighted sum of the inputs (total net input). Activation functions with a bounded range are often termed squashing functions. The output a of this transfer function is binary depending on whether the input meets a specified threshold, T:

$$a = f\left(\sum_{i=0}^{n} w_i x_i - T \right)$$

(1.2)

If the total net input is less than 0, then the output of the neuron is 0, otherwise it is 1 (Duch and Jankowski, 1999). The choice of transfer function strongly influences the complexity and performance of neural networks and may be a linear or nonlinear function of n (Hagen et al., 1996). The simplest squashing function is a step function (Figure 1.4):

$$a = \begin{cases} 0 \,...\, \text{if} \sum_{i=0}^{n} w_i x_i \rhd T \\ \\ 1 \,...\, \text{if} \sum_{i=0}^{n} w_i x_i \lhd T \end{cases}$$

(1.3)

This function is used in perceptrons (see Chapter 2, Section 2.2.1) and related neural network models. Documented advantages of this approach are the high

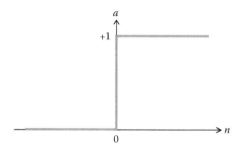

FIGURE 1.4   Schematic of a step function. The output is a_1 if the input sum is above a certain threshold and a_0 if the input sum is below a certain threshold.

speed of associated computations and easy realization in the hardware (Duch and Jankowski, 1999). Yet, it is limited to use in perceptrons with a single layer of neurons.

Figure 1.5 displays three additional transfer functions commonly employed in neural networks to generate output. The linear transfer function, illustrated in Figure 1.5a, has an output that is equal to its input [$a = purelin\ (n)$]. Neural networks similar to perceptrons, but with linear transfer functions, are termed linear filters. This function computes a neuron's output by merely returning the value passed directly to it. Hence, a linear network cannot perform a nonlinear computation. Multilayer perceptrons using a linear transfer function have equivalent single-layer networks; a nonlinear function is therefore necessary to gain the advantages of a multilayer network (Harrington, 1993).

Nonlinear transfer functions between layers permit multiple layers to deliver new modeling capabilities for a wider range of applications. Log-sigmoid transfer functions (Figure 1.5b) take given inputs and generate outputs between 0 and 1 as the neuron's net input goes from negative to positive infinity [$a = \log\ sig\ (n)$], and is mathematically expressed as

$$a = \frac{1}{1+e^{-n}}$$ (1.4)

This function is commonly used in back-propagation networks, in large part because it is differentiable (Harrington, 1993; Hagan et al., 1996). Alternatively, multilayer networks may use Gaussian-type functions or the hyperbolic tan-sigmoid transfer function [$a = \tan\ sig\ (n)$]. Gaussian-type functions are employed in radial basis function networks (Chapter 2, Section 2.2.2) and are frequently used to perform function approximation. The hyperbolic tan-sigmoid function is shown in Figure 1.5c and represented mathematically as

$$a = \frac{e^n - e^{-n}}{e^n + e^{-n}}$$ (1.5)

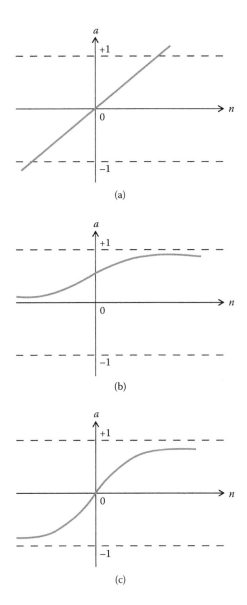

FIGURE 1.5 Three common transfer functions employed in neural networks to generate outputs: (a) linear transfer function [$a = purelin\ (n)$], (b) log-sigmoid transfer function [$a = \log sig\ (n)$], and (c) hyperbolic tangent-sigmoid transfer function [$a = \tan sig\ (n)$].

The hyperbolic tangent is similar to the log-sigmoid but can exhibit different learning dynamics during the training phase. The purpose of the sigmoid function is to generate a degree of nonlinearity between the neuron's input and output. Models using sigmoid transfer functions often display enhanced generalized learning characteristics and produce models with improved accuracy. One potential drawback is the propensity for increased training times.

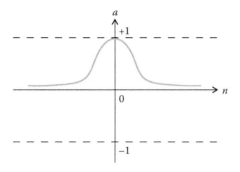

FIGURE 1.6   Schematic of the radial basis function (RBF). RBFs [a = radbas (n)] have radial symmetry about a center and have a maximum of 1 when their input is 0. They are further characterized by a localization property (center) and activation hypersurface (a hyperellipsoid in general cases and hypersphere when the covariance matrix is diagonal).

Investigators are now looking beyond these commonly used functions as there is a growing understanding that the choice of transfer function is as important as the network architecture and learning algorithms. For example, radial basis functions (RBFs), real-valued functions whose value depends only on the distance from the origin, are increasingly used in modern applications (Corsini et al., 2003). In contrast to sigmoid functions, radial basis functions [a = radbas (n)] have radial symmetry about a center and have a maximum of 1 when their input is 0. Typically, radial basis functions are assumed to be Gaussian-shaped (Figure 1.6) with their values decreasing monotonically with the distance between the input vector and the center of each function (Corsini et al., 2003). The Gaussian function is given by

$$\phi(r) = \exp\left(\frac{-r^2}{\beta^2}\right)$$ (1.6)

Others include the thin-spline-function:

$$\phi(r) = r^2 \log(r)$$ (1.7)

the multiquadric function:

$$\phi(r) = (r^2 + \beta^2)^{1/2}$$ (1.8)

and the inverse multiquadric function:

$$\phi(r) = \left(\frac{1}{(r^2 + \beta^2)^{1/2}}\right)$$ (1.9)

where r = Euclidean distance between an associated center, c_j, and the data points, and β = a real variable to be decided by users. A set of radial basis functions are employed to construct function approximations of the form (Chen et al., 2008):

$$F(\mathbf{x}) = \sum_{j=1}^{n_c} w_i \phi \left(\|\mathbf{x}_i - \mathbf{c}_j\| \right) + w_0 \qquad (1.10)$$

where $F(\mathbf{x})$ = the approximating function represented as a sum of N radial basis functions, each associated with different centers c_j, and weighted by an appropriate coefficient w_i. The term w_0 = a constant term that acts as a shift in the output level, and x_i = the input or the pattern vector. Note that $\| x \|$ denotes the norm, which is usually taken to be Euclidean. As will be discussed in Chapter 2, RBFs are embedded in two-layer neural networks, where each hidden unit implements a radial activated function (Bors and Pitas, 1996).

A large number of alternative transfer functions have been proposed and exploited in modern research efforts. Universal transfer functions, parameterized to change from localized to a delocalized type, are of greatest interest (Duch and Jankowski, 1999). For example, Hoffmann (2004) discussed the development of universal basis functions (UBFs) with flexible activation functions parameterized to change their shape smoothly from one functional form to another. This allows the coverage of bounded and unbounded subspaces depending on the data distribution. UBFs have been shown to produce parsimonious models that tend to generalize more efficiently than comparable approaches (Hoffmann, 2004). Other types of neural transfer functions being considered include functions with activations based on non-Euclidean distance measures, bicentral functions, biradial functions formed from products or linear combinations of pairs of sigmoids, and extensions of such functions making rotations of localized decision borders in highly dimensional spaces practical (Duch and Jankowski, 1999). In summary, a variety of activation functions are used to control the amplitude of the output of the neuron. Chapter 2 will extend the discussion on artificial neuron models, including network connectivity and architecture considerations.

1.3   NEURAL NETWORK APPLICATION AREAS

Neural networks are nonlinear mapping structures shown to be universal and highly flexible junction approximators to data-generating processes. Therefore, they offer great diversity in the type of applications in which neural networks can be utilized, especially when the underlying data-generating processes are unknown. Common neural network applications include those used in the following activities:

1. Prediction and forecasting
2. System identification and process control
3. Classification, including pattern recognition
4. Optimization
5. Decision support

The analysis of biological and environmental data is inherently complex, with data sets often containing nonlinearities; temporal, spatial, and seasonal trends; and non-Gaussian distributions. The ability to forecast and predict values of time-sequenced data will thus go a long way in impacting decision support systems. Neural network architectures provide a powerful inference engine for regression analysis, which stems from the ability of neural networks to map nonlinear relationships, that is more difficult and less successful when using conventional time-series analysis (May et al., 2009). Neural networks provide a model of the form (May et al., 2009):

$$y = F(x) + \varepsilon \tag{1.11}$$

where F is an estimate of some variable of interest, y, $x = x_1,...,x_n$ and denotes the set of input variables or predictors, and ε is noise or an error term. The training of the neural network is analogous to parameter estimation in regression. As discussed in Section 1.2, neural networks can approximate any functional behavior, without the prerequisite a priori knowledge of the structure of the relationships that are described. As a result, numerous applications of predictive neural network models to environmental and biological analyses have been reported in the literature. For example, neural networks have been incorporated into urban air quality studies for accurate prediction of average particulate matter ($PM_{2.5}$ and PM_{10}) concentrations in order to assess the impact of such matter on the health and welfare of human populations (Perez and Reyes, 2006; Dong et al., 2009). They have also been widely incorporated into biological studies, for example, from the use of neural networks in predicting the reversed-phase liquid chromatography retention times of peptides enzymatically digested from proteome-wide proteins (Petritis et al., 2003), to the diagnosis of heart disease through neural network ensembles (Das et al., 2009).

Neural networks have also been widely accepted for use in system identification and process control, especially when complex nonlinear phenomena are involved. They are used in industrial processes that cannot be completely identified or modeled using reduced-order linear models. With neural networks, empirical knowledge of control operations can be learned. Consider a wastewater treatment process. Increased concentrations of metals in water being discharged from a manufacturing facility could indicate a problem in the wastewater treatment process. Neural networks as statistical process control can identify shifts in the values monitored, leading to early detection of problems and appropriate remedial action (Cook et al., 2006). Other representative applications in process control include fouling control in biomass boilers (Romeo and Gareta, 2009) and control of coagulation processes in drinking water treatment plants (Bloch and Denoeux, 2003).

Pattern recognition techniques seek to identify similarities and regularities present in a given data set to achieve natural classification or groupings. Reliable parameter identification is critical for ensuring the accuracy and reliability of models used to assess complex data sets such as those acquired when studying natural systems. Neural networks lend themselves well to capturing the relationships and interactions among input variables when compared to traditional approaches such as generalized logistic models (GLM). As a result, neural network models have been routinely incorporated into modern environmental modeling efforts. For example, a neural network approach was utilized in modeling complex responses of shallow lakes using carp biomass, amplitude of water levels fluctuations, water levels, and a morphology index as input parameters (Tan and Beklioglu, 2006). Inherent complexities (e.g., nonlinearities) of ecological process and related interactions were overcome by the use of neural networks. Predictions in explaining the probability of submerged plant occurrences were in strong agreement with direct field observations.

Information systems and technology are an integral part of biological and environmental decision-making processes. Effective management of water resources, for example, relies on information from a myriad of sources, including monitoring data, data analysis, and predictive models. Stakeholders, regulatory agencies, and community leaders of various technical backgrounds and abilities need to be able to transform data into usable information to enhance understanding and decision making in water resource management. In such a situation, models would be able to reproduce historical water use trends and generate alternative scenarios of interest to affected communities, and aid in achieving water quality management objectives. Neural networks can also provide clinicians and pharmaceutical researchers with cost-effective, user-friendly, and timely analysis tools for predicting blood concentration ranges in human subjects. This type of application has obvious health-related benefits and will likely provide a clinically based decision support system for clinicians and researchers to follow and direct appropriate medical actions.

1.4   CONCLUDING REMARKS

The importance of developing and applying neural network techniques to further our understanding of complex biological and environmental processes is evident. The efficacy of artificial neural network models lies in the fact that they can be used to infer a function from a given set of observations. Although the broad range of applicability of neural networks has been established, new and more efficient models are in demand to meet the data-rich needs of modern research and development. Subsequent chapters will provide content related to network architecture, learning paradigms, model selection, sensitivity analysis, and validation. Extended considerations in data collection and normalization, experimental design, and interpretation of data sets will be provided. Finally, theoretical concepts will be strengthened by the addition of modern research applications in biological and environmental analysis efforts from global professionals active in the field.

REFERENCES

Bennett, M., Contreras, J., Bukauskas, F., and Sáez, J. 2003. New roles for astrocytes: Gap junction hemichannels have something to communicate. *Trends in Neuroscience* 26: 610–617.

Bloch, G., and Denoeux, T. 2003. Neural networks for process control and optimization: Two industrial applications. *ISA Transactions* 42: 39–51.

Bobrow, D.G., and Brady, J.M. 1998. Artificial intelligence 40 years later. *Artificial Intelligence* 103: 1–4.

Bors, A.G., and Pitas, I. 1996. Median radial basis function neural network. *IEEE Transactions on Neural Networks* 7: 1351–1364.

Chalmers, D. 1996. *The Conscious Mind: In Search of a Fundamental Theory*. Oxford: Oxford University Press.

Chen, S., Cowan, C.F.N., and Grant, P.M. 2008. Orthogonal least squares learning for radial basis function networks. *IEEE Transactions on Neural Networks* 2: 302–309.

Churchland, P.M. 1989. *A Neurocomputational Perspective*. MIT Press: Cambridge, MA.

Cook, D.F., Zobel, C.W., and Wolfe, M.L. 2006. Environmental statistical process control using an augmented neural network classification approach. *European Journal of Operational Research* 174: 1631–1642.

Copeland, B.J., and Proudfoot, D. 2000. What Turing did after he invented the universal Turing machine. *Journal of Logic, Language and Information* 9: 491–509.

Corsini, G., Diani, M., Grasso, R., De Martino, M., Mantero, P., and Serpico, S.B. 2003. Radial bias function and multilayer perceptron neural networks for sea water optically active parameter estimation in case II waters: A comparison. *International Journal of Remote Sensing* 24: 3917–3932.

Cowan, J.D. 1990. Discussion: McCulloch-Pitts and related neural nets from 1943 to 1989. *Bulletin of Mathematical Biology* 52: 73–97.

Das, R., Turkoglu, I., and Sengur, A. 2009. Effective diagnosis of heart disease through neural networks ensembles. *Expert Systems with Applications* 36: 7675–7680.

Dong, M., Yong, D., Kuang, Y., He, D., Erdal, S., and Kenski, D. 2009. $PM_{2.5}$ concentration prediction using hidden semi-Markov model-based times series data mining. *Expert Systems with Applications* 36: 9046–9055.

Dubois, D. and Prade, H. 2004. On the use of aggregation operations in information fusion processes. *Fuzzy Sets and Systems* 142: 143–161.

Duch, W., and Jankowski, N. 1999. Survey of neural transfer functions. *Neural Computing Surveys* 2: 163–212.

Edelman, G.M. 1992. *Bright Air, Brilliant Fire: On the Matter of the Mind*. Basic Books: New York.

Fodor, J.A. 1975. *The Language of Thought*. Harvard University Press: Cambridge, MA.

Freeman, W.J. 2001. *How Brains Make Up Their Minds*. Columbia University Press: New York.

Gallistel, C.R., and Gibbon, J. 2002. *The Symbolic Foundations of Conditioned Behavior*. Lawrence Erlbaum Associates: Mahwah, NJ.

Giuliodori, M.J., and Zuccolilli, G. 2004. Postsynaptic potential summation and action post initiation: Function following form. *Advances in Physiology Education* 28: 79–80.

Hagan, M.T., Demuth, H.B., and Beale, M.H. 1996. *Neural Network Design*. PWS Publishing Company: Boston.

Hanrahan, G. 2010. Computational neural networks driving complex analytical problem solving. *Analytical Chemistry*, 82: 4307–4313.

Harrington, P.B. 1993. Sigmoid transfer functions in backpropagation neural networks. *Analytical Chemistry* 65: 2167–2168.

Hebb, D.O. 1949. *The Organization of Behaviour*. John Wiley & Sons: New York.

Hoffmann, G.A. 2004. *Transfer Functions in Radial Basis Function (RBF) Networks in Computational Science—ICCS 2004*. Springer Berlin/Heidelberg, pp. 682–686.

Hopfield, J. J. 1982. Neural networks and physical systems with emergent collective computational abilities. *Proceedings of the National Academies of Sciences* 79: 2554–2558.

Hopfield, J., and Tank, D. W. 1985. "Neural" computation of decisions in optimization problems. *Biological Cybernetics* 55: 141–146.

Jalali-Heravi, M. 2008. Neural networks in analytical chemistry. In D.S. Livingstone (Ed.), *Artificial Neural Networks: Methods and Protocols*. Humana Press: New Jersey.

Johannet, A., Vayssade, B., and Bertin, D. 2007. Neural networks: From black box towards transparent box. Application to evapotranspiration modeling. *Proceedings of the World Academy of Science, Engineering and Technology* 24: 162–169.

Kock, C. 1999. *Biophysics of Computation: Information Processing in Single Neurons.* Oxford University Press: Oxford, UK.

Konstantinos, P., Kangas, L.J., Ferguson, P.L., Anderson, G.A., Paša-Tolić, L., Lipton, M.S., Auberry, K.J., Strittmatter, E.F., Shen, Y., Zhao, R., and Smith, R.D. 2003. *Analytical Chemistry* 75: 1039–1048.

Luger, G.F., and Stubblefield, W.A. 1997. *Artificial Intelligence: Structures and Strategies for Complex Problem Solving*, 3rd Edition. Addison-Wesley Longman: Reading, MA.

Marinia, F., Bucci, R., Magrí, A.L., and Magrí, A.D. 2008. Artificial neural networks in chemometrics: History, examples and perspectives. *Microchemical Journal* 88: 178–185.

May, R.J., Maier, H.R., and Dandy, G.C. 2009. Developing artificial neural networks for water quality modelling and analysis. In G. Hanrahan, Ed., *Modelling of Pollutants in Complex Environmental Systems*. ILM Publications: St. Albans, U.K.

McCulloch, W., and Pitts, W. 1943. A logical calculus of the ideas immanent in nervous activity. *Bulletin of Mathematics and Biophysics* 5: 115–133.

Minsky, M.L., and Papert, S.A. 1969. *Perceptrons: An Introduction to Computational Geometry*. MIT Press: Cambridge, MA.

Mira, J. 2008. Symbols versus connections: 50 years of artificial intelligence. *Neurocomputing* 71: 671–680.

Olden, J.D., and Jackson, D.A. 2002. Illuminating the "black box": A randomization approach for understanding variable contributions in artificial neural networks. *Ecological Modelling* 154: 135–150.

Oussar, Y., and Dreyfus, G. 2001. How to be a gray box: Dynamic semi-physical modelling. *Neural Networks* 14: 1161–1172.

Özesmi, S.L., Tan, C.O., and Özesmi, U. 2006. Methodological issues in building, training, and testing artificial neural networks in ecological applications. *Ecological Modelling* 195: 83–93.

Parker, X., and Newsom, X. 1998. Sense and the single neuron: Probing the physiology of perception. *Annual Review of Neuroscience* 21: 227–277.

Paulsen, O., and Sejnowski, T.J. 2000. Natural patterns of activity and long term synaptic plasticity. *Current Opinions in Neurobiology* 10: 172–179.

Perea, G., and Araque, A. 2002. Communication between astrocytes and neurons: a complex language. *Journal of Physiology-Paris* 96: 199–207.

Perez, P., and Reyes, J. 2006. An integrated neural network model for PM_{10} forecasting. *Atmospheric Environment* 40: 2845–2851.

Peters, J., Niko, E.C., Verhoest, R.S., Van Meirvenne, M., and De Baets, B. 2009. Uncertainty propagation in vegetation distribution models based on ensemble classifiers. *Ecological Modelling* 220: 791–804.

Petritis, K., Kangas, L.J., Ferguson, P.L., Anderson, G.A., Paša-Tolić, L., Lipton, M.S., Auberry, K.J., Strittmatter, E.F., Shen, Y., Zhao, R. and Smith, R.D. 2003. Use of artificial neural networks for the accurate prediction of peptide liquid chromatography elution times in proteome analyses. *Analytical Chemistry* 75: 1039–1048.

Piccinini, G. 2004. The first computational theory of mind and brain: A close look at McCulloch and Pitt's "logical calculus of ideas immanent in nervous activity." *Synthese* 141:175–215.

Piccinini, G. 2008. Some neural networks compute, others don't. *Neural Networks* 21: 311–321.

Recknagel, F., French, M., Harkonen, P., and Yabunaka, K. 1997. Artificial neural network approach for modelling and prediction of algal blooms. *Ecological Modelling* 96: 11–28.

Rejyol, Y., Lim, P., Belaud, A., and Lek, S. 2001. Modelling of microhabitat used by fish in natural and regulated flows in the river Garonne (France). *Ecological Modelling* 146: 131–142.

Rumelhart, D.E., and McClelland, J.L. 1986. *Parallel Distributed Processing: Explorations in the Microstructure of Cognition*. MIT Press: Cambridge, MA.

Scardi, M. 2001. Advances in neural network modelling of phytoplankton primary production. *Ecological Modelling* 146: 33–45.

Shagrir, O. 2006. Why we view the brain as a computer. *Synthese* 153: 393–416.

Smolensky, D. 1987. Connectionist AI, Symbolic AI, and the Brain. *Artificial Intelligence Review* 1: 95–109.

Spector, L. 2006. Evolution of artificial intelligence. *Artificial Intelligence* 170: 1251–1253.

Tan, C.O., and Beklioglu, M. 2006. Modeling complex nonlinear responses of shallow lakes to fish and hydrology using artificial neural networks. *Ecological Modelling* 196: 183–194.

Widrow, B., and Hoff, M.E. 1960. Adaptive switching circuits. *IRE WESCON Convention Record* 4: 96–104.

Xing, H., Huang, S.H., and Shi, J. 2003. Rapid development of knowledge-based systems via integrated knowledge acquisition. *Artificial Intelligence for Engineering Design, Analysis and Manufacturing* 17: 221–234.

2 Network Architectures

2.1 NEURAL NETWORK CONNECTIVITY AND LAYER ARRANGEMENT

Details of the elemental building blocks of a neural network, for example, individual neurons, nodal connections, and the transfer functions of nodes, were provided in Chapter 1. Nonetheless, in order to fully understand the operation of the neural network model, knowledge of neuron connectivity and layer arrangement is essential. Connectivity refers to the level of interaction within a system; in neural network terms, it refers to the structure of the weights within the networked system. The selection of the "correct" interaction is a revolving, open-ended issue in neural network design and is by no means a simple task. As will be presented in subsequent chapters, there are countless methods used to aid in this process, including simultaneous weight and structure updating during the training phase and the use of evolutionary strategies: stochastic techniques capable of evolving both the connection scheme and the network weights. Layer arrangement denotes a group of neurons that have specialized function and are largely processed through the system as a collective. The ability to interpret and logically assemble ways in which neurons are interconnected to form the networks or network architectures would thus prove constructive in model development and final application. This chapter will provide a solid background for the remainder of this book, especially Chapter 3, given that training of neural networks is discussed in detail.

2.2 FEEDFORWARD NEURAL NETWORKS

Feedforward neural networks are arguably the simplest type of artificial neural networks and characterized as network connections between the units do not form a directed cycle (Agatonivic-Kustrin and Beresford, 2000). Information proceeds in one direction: forward progress from the input layer and on to output layer. The activity of the input layers represents the data that are fed into individual networks. Every input neuron represents some independent variable that has an influence over the output of the neural network. As will be discussed in Section 2.2.1, activities of hidden layers are determined by the activities of the input units and the weights on the connections between the input and the hidden units.

2.2.1 THE PERCEPTRON REVISITED

As discussed in Chapter 1, an artificial neuron can receive excitatory or inhibitory inputs similar to its biological counterpart. And as was shown in Chapter 1,

Figure 1.3, a modified Rosenblatt's model of a neuron depicts the simplicity in structure, yet this neuron displays a great deal of computing potential within its boundaries. Even so, the perceptron is limited in terms of application as it only generates a binary output with fixed weight and threshold values. By connecting multiple neurons feeding forward to one output layer, the true computing power of neural networks comes to light. Consider the single-layer feedforward network (perceptron) eloquently discussed in detail by Hagan et al. (1996) and shown in generalized format in Figure 2.1. Included in single-layer perceptrons are the following key components: weight matrix **W**, summers, the bias vector **b**, transfer function boxes, and the output vector **a**. Each element in the input vector **x** is connected to each neuron through the weight matrix **W** (Hagan et al. [1996]):

$$\mathbf{W} = \begin{bmatrix} w_{1,1} & w_{1,2} \cdots w_{1,R} \\ w_{2,1} & w_{2,2} \cdots w_{2,R} \\ \vdots & \vdots \quad \vdots \\ w_{S,1} & w_{S,2} \cdots w_{S,R} \end{bmatrix} \tag{2.1}$$

As represented in the preceding equation, the row of indices of the elements of matrix **W** indicates the destination neuron associated with that weight. In contrast, column indices indicate the source of the input for that particular weight. Although other mathematical operations are encountered in the study and in the application of neural networks, matrix operations are specifically used in neural programming to both train the network and calculate relevant outputs. While it is beyond the

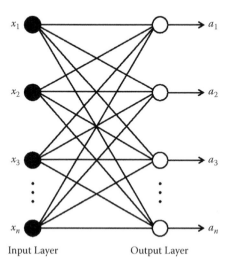

FIGURE 2.1 Schematic of a representative single-layer perceptron with one input and one output layer of processing units. Inputs and connection weights embedded within a perceptron's structure are typically real values, both positive and negative.

scope of this book to formally cover such topics, a review of basic matrix notation operations can be found in Appendix I. Those wishing to gain a more detailed
understanding are encouraged to consult Haton's book titled *Introduction to Neural
Networks for Java.*

Single-layer perceptrons are straightforward to set up, train, and explicitly link
to statistical models, that is, sigmoid output functions, thus allowing a link to posterior probabilities. Nonetheless, single-layer perceptrons do have their limitations.
Consider the classic example of the inability of single-layer perceptrons in solving
the XOR binary function (Minsky and Papert, 1969). It is easier to visualize this
problem if neural network decision boundaries between classes are reviewed in
detail (Fausett, 1994). For Figure 2.2a, the output is given by

$$out = \text{sgn}(w_1 x_1 + w_2 x_2 - \theta) \tag{2.2}$$

The decision boundary that is between $out = 0$ and $out = 1$ is given by

$$w_1 x_1 + w_2 x_2 - \theta = 0 \tag{2.3}$$

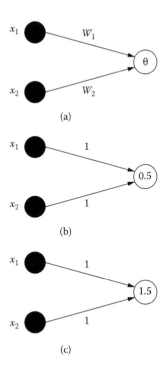

FIGURE 2.2 (a) A single-layer perceptron made up of two inputs connected to a single
output unit. The output of the network is determined by calculating a weighted sum of its
two inputs and comparing this value with a threshold. If the net input (net) is greater than the
threshold, the output is 1, otherwise it is 0. Visual inspection of weights and thresholds for (b)
logical OR and (c) logical AND.

that is, along a straight line, it can be shown that

$$x_2 = \left(\frac{-w_1}{w_2} \right) x_1 + \left(\frac{\theta}{w_2} \right)$$ (2.4)

Given Equation 2.4, the decision boundaries will always be straight lines in two-dimensional representations (see the following discussion).

One can construct simple networks that perform NOT, AND, and OR operations, and any logical function possible can be constructed from these three operations (Widrow and Lehr, 1990). In order to implement these operations, careful determination of the weights and thresholds is needed. For example, visual inspection of weights and thresholds for OR and AND is shown in Figures 2.2b and 2.2c, respectively. The decision boundaries for AND and OR can be plotted as shown in Figures 2.3a and 2.3b, respectively. For both plots, the two axes are the inputs, which can take the value of either 0 or 1. The numbers on the graph are the expected output for a given input. Here, the single-layer perceptron can perform the AND and OR functions. For each case, linear separation is achieved, and the perceptron algorithm converges and positions the decision surface in the form of a hyperplane between the designated classes. The proof of convergence is known as the perceptron convergence theorem. The basic premise behind the proof is to acquire upper and lower bounds on the span of the weight vector; if the span is finite, then the perceptron has in fact converged (this implies that the weights have in fact changed a finite number of times).

The XOR operator discussed earlier is not linear separable and cannot be achieved by a single perceptron. Its truth table is given as follows:

x_1	x_2	Output
0	0	0
0	1	1
1	0	1
1	1	0

The resultant output belongs to either of two classes (0 or 1) because the two classes are nonseparable. In order to solve the XOR function, two straight lines to separate the differing outputs are needed (Figure 2.4).

It was Minsky and Papert (1969) who first offered a solution to the XOR problem, by combining perceptron unit responses using a second layer of units. Multilayer perceptrons (MLPs) consist of an input layer, one or more hidden layers, and an output layer, and can ultimately be used to approximate any continuous function (Zhou et al., 2008). A generalized structure of an MLP is presented in Figure 2.5. Close examination reveals one neuron in the input layer for each predictor variable. If categorical variables are considered, $N - 1$ neurons are used to signify the N categories of the variable. The input layer standardizes the predictor variable values $(x_1, ..., x_n)$ to a range of −1 to 1. This layer distributes the values to each of the neurons in the hidden layer. In addition to the predictor variables, there is a constant input of 1.0 (bias) that is fed to the hidden layer. The bias is multiplied by a weight and added

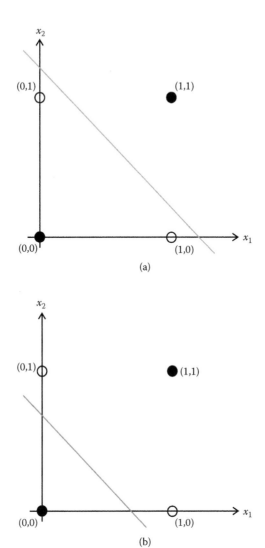

(a)

(b)

FIGURE 2.3 Decision boundaries for (a) AND and (b) OR. For the plots, the two axes are the inputs, which can take a value of either 0 or 1. The numbers on the graph are the expected output for a given input.

to the sum going into the neuron to generate appropriate output (see Figure 1.3 and caption description).

In the hidden layer, the value from each input neuron is multiplied by a given weight, and the resulting weighted values are summed appropriately. The weighted sum is then fed into a selected transfer function with final distribution to the output layer. The determination of the number of hidden layers is an important consideration for practitioners. If one layer is chosen, the model can approximate arbitrarily for any functions that contain a continuous mapping from one finite space to another.

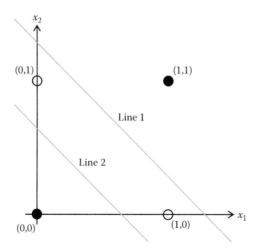

FIGURE 2.4 The use of multilayer perceptrons (MLP) to solve the XOR function. Multiple hyperplanes can be drawn to separate the patterns with +1 desired outputs from ones with −1 desired outputs.

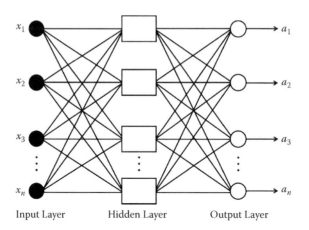

FIGURE 2.5 A generalized structure of a MLP consisting of an input layer, one or more hidden layers, and an output layer. The output of each layer is connected to the input of nodes in the subsequent layer. Inputs of the first layer are actual inputs to the neural network, while the last layer forms the output of the network. (From Hanrahan, G. 2010. *Analytical Chemistry*, 82: 4307–4313. With permission from the American Chemical Society.)

Models with two hidden layers result in the representation of an arbitrary decision boundary with arbitrary accuracy with rational activation functions. In addition, models with two hidden layers have been shown to approximate any smooth mapping with high accuracy (Priddy and Keller, 2005). Note that practitioners must also determine how many neurons will be present in each of the chosen hidden layers. The number of output neurons should directly relate to the type of task that the neural network is to perform. For example, if the neural network is to be used to classify items into groups, it is then customary to have one output neuron for each group that the item is to be assigned into. Finally, the value from each hidden layer neuron is multiplied by a weight, and the resulting weighted values are added together, producing a combined value. The weighted sum is fed into a selected transfer function, which outputs a value, where the values designated by the character a are the outputs of the neural network.

MLPs are arguably the most popular among currently available neural techniques. They have been shown to be especially useful in classification problems with the use of sigmoidal transfer functions providing soft hyperplanes dividing the input space into separate regions (Feng and Hong, 2009). Classification itself is a very common task in modern research, with a variety of methods reported in the literature, including the use of neural networks. Zomer (2004) reports that the choice of the most appropriate method depends on the specific nature of the problem, for example, how many classes, objects, and variables describe the data set; what is the expected complexity of the boundaries between the classes (e.g., from exploratory analysis); what is the noise level; and the fraction of outliers in the data. Duch et al. (1999) compared MLPs to statistical discriminant techniques, but stressed that the combination of soft sigmoids allows for representation of more complex, nonlinear decision boundaries. They also cautioned that this becomes a potential weakness of MLPs when sharp decision borders are compulsory. The authors used the example of classification borders conforming to a simple logical rule $x_1 > 1 \land x_2 > 1$ and stated that such borders are easily represented by two hyperplanes. However, there is no way to represent them accurately using the soft sigmoidal functions typically used in MLPs. A more recent examination of classification methodologies of multilayer perceptrons with sigmoid activation functions was performed by Daqi and Yan (2005). They showed that in the input space the hyperplanes determined by the hidden basis functions with values of 0 did not play the role of decision boundaries, and such hyperplanes certainly did not go through the marginal regions between different classes. For an extended discussion on the overall classification error of MLPs, consult the work of Feng and Hong (2009).

Despite the foregoing listed weaknesses, MLPs have been widely used in classifying biological and environmental systems. For example, work by Gardner and Dorling (1998) extensively reviewed the use of the MLP in the atmospheric sciences. They first provided a detailed overview of MLP concepts and training procedures, and subsequently turned their attention to reviewing example applications in the field. This was followed by a discussion of common practical problems and limitations associated with the neural network approach. More recent sources have covered biological and environmental systems study areas in greater detail (Yuste and Dorado, 2006; Seidel et al., 2007; Jalali-Heravi, 2008; May et al., 2009). MLPs have also been widely applied to general regression, optimization, automatic control, and pattern completion problems (Mielniczuk and Tyrcha, 1993; García-Altés, 2007).

2.2.2 RADIAL BASIS FUNCTION NEURAL NETWORKS

With the development and use of radial basis functions (RBFs) came radial basis function neural networks (RBFNNs) in the late 1980s, which now offer unique

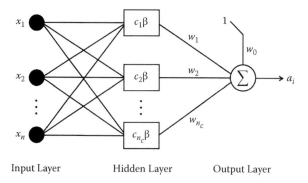

Input Layer   Hidden Layer   Output Layer

FIGURE 2.6   Basic structure of a radial basis function neural network displaying a feedforward network with three layers, with the output a linear combination of the output from its hidden units. Each hidden unit implements a radial basis function (RBF).

advantages when compared with MLPs. RBFNNs demonstrate good generalization ability with a simple network structure that avoids lengthy calculations. The basic structure (Figure 2.6) is a feedforward network with three layers, with an overall output being a linear combination of the output from its hidden units (with each hidden unit implementing a radial basis function):

$$a_i = \sum_{j=1}^{n_c} w_i \, \phi(\|\mathbf{x}_i - \mathbf{c}_j\|) + w_0 \tag{2.5}$$

In pattern classification applications, the aforementioned Gaussian function (Chapter 1, Equation 1.6) is commonly preferred. The network inputs represent feature entries, with each output corresponding to a class. RBFNNs separate class distributions by local Gaussian kernel functions. Studies have shown that a single kernel function will not give particularly good representations of the class-conditional distributions (Bishop, 2005). The Gaussian kernel for dimensions greater than one, for example, the term N, can be described as a regular product of N one-dimensional kernels. Given that higher-dimensional Gaussian kernels are regular products of one-dimensional Gaussians, they are labeled separable. As a result, using a separate mixture model to represent each of the conditional densities has been applied in many routine and experimental applications. In comparison, a multilayer perceptron separates classes by hidden units that form hyperplanes in the input space. If time series modeling is of interest, then thin-plate-spline functions (Chapter 1, Equation 1.7) have proved moderately useful. Here, the network inputs represent data samples at certain past time-laps, with the network demonstrating only one output (Chen et al., 1991).

RBFNNs have also been shown to implement Bayesian frameworks. In the Bayesian approach, a portion of the prior knowledge is specified more explicitly, in the form of prior distributions for the model parameters, and hyperpriors for the parameters of the prior distributions (Lampinen and Vehtar, 2001). Regardless of the sample size, acquired results can subsequently be used to compute posterior probabilities of original hypotheses based on the data available. Whether used separately or in conjunction with classical methods, Bayesian methods provide modelers with a formal set of tools for analyzing complex sets of data. Distinct advantages of such techniques include

1. Suitability of use when missing data is evident.
2. They allow a combination of data with domain knowledge.
3. The facilitation of learning about casual relationships between variables.
4. Providing a method for avoiding overfitting of data.
5. Good prediction accuracy regardless of sample size.
6. Can be combined with decision analytic tools for aid in management-related activities.

When considering complex models such as neural networks, the relationship between the actual domain knowledge of the experts and the priors for the model parameters is not simple, and thus it may in practice be difficult to incorporate very sophisticated background information into the models via the priors of the parameters (Lampinen and Vehtar, 2001). Despite this possibility, the Bayesian framework offers an integrated approach for the development of minimal neural network models. This framework allows the encapsulation of combined uncertainty induced by the data in all parameters, including variances and covariances, and acceptably propagates this through the model arrangement. For example, Andrieu et al. (2001) presented a hierarchical full Bayesian model for RBFNNs using a reversible-jump Markov chain Monte Carlo (MCMC) method as computation. Results obtained using this approach were superior to those previously reported and robust with respect to prior specification. Ultimately, RBFFN performance is attributed to the number and positions of the radial basis functions, their overall shape, and the method used in the learning process for input-output mapping.

Normalized radial basis function neural networks (NRBFNNs) possess novel computational characteristics due to minor modification of their equation through a normalizing factor:

$$\phi(\mathbf{x}) = \frac{\phi(\|\mathbf{x} - \mathbf{c}_i\|)}{\sum_{j=1}^{M} \phi(\|\mathbf{x} - \mathbf{c}_j\|)} \tag{2.6}$$

where M = the total number of kernels. In NRBFNNs, the traditional roles of weights and activities in the hidden layer are switched. As reported by Bugmann (1998), the hidden nodes perform a function similar to a Voronoi tessellation of the input space, and the output weights become the network's output over the partition defined by the hidden nodes. Considering a set of N points in a given plane, Voronoi tessellation partitions the domain into polygonal regions, the boundaries of which are the perpendicular bisectors of the lines combining the points (Zhu, 2006). Upon comparison with RBFNNs, Bugmann went on to report that NRBFNNs have superior generalization properties that are beneficial in complex classification tasks. This is attributable to the property of NRBFNNs to produce a significant output even for input vectors far from the center of the receptive field of any of the hidden nodes (Bugmann, 1998).

Probabilistic neural networks (PNNs) and generalized regression neural networks (GRNNs) are specialized variants of the more general RBFNN. Both have been particularly useful in classification-type applications. PNNs, first reported in 1990 by Specht, operate on the concept of the Parzen windows classifier and its application to Bayesian statistics (Specht, 1990). The Parzen windows classifier is a nonparametric procedure where the joint probability density function (PDF) of a given input vector and a certain input class is approximated by the superposition of a number of (usually Gaussian) functions or windows (Gelman et al., 2003). In contrast to conventional back-propagation neural networks and RBFNNs, no learning rule is required, no weights are assigned to the links connecting the layers, and no predefined convergence criteria are needed (Adeli and Panakkat, 2009). GRNNs were also developed by Specht (1991) with subsequent modification by Schiøler and Hartmann (1992). They are based on established statistical principles and converge to an optimal regression surface through the use of the previously discussed Parzen windows classifier. A more recent modification by Tomandl and Schober (2001) presented suitable model-free approximations of the mapping between independent and dependent variables.

2.3   RECURRENT NEURAL NETWORKS

In contrast to the strictly feedforward approaches described earlier, recurrent neural networks (RNNs) have at least one feedback (closed loop) connection present. Ostensibly, such networks have activation feedback, which symbolizes short-term memory. As a result, RNNs are able to perform sequence recognition, sequence reproduction, and temporal association activities (Botvinick and Plaut, 2006). RNNs take on two types of feedback connections: (1) local feedback (linkages that pass the output of a given neuron to itself) and (2) global feedback (linkages that pass the output of a neuron to other neurons in the same or lower layers). In the latter connection scheme, RNNs can take on the characteristics of a multilayer perceptron, with one critical exception: the output of one layer can route back to the input layer of a previous layer (Figure 2.7).

Feedback in RNNs is modified by a set of weights so as to enable automatic adaptation through a learning process (e.g., back-propagation). A state layer is updated not only with the external input of the network but also with activation from the previous forward-propagation process. Internal states retain previous information and use this historical record or "memory" when functioning under new inputs. Interest in such memories has been spurred by the influential work of Hopfield in the early 1980s, who demonstrated how a simple discrete nonlinear dynamical system can exhibit associative recall of stored binary patterns through collective computing (Hassoun, 1993). There are two main requirements for associative memory: (1) every memory should be in equilibrium with the network, and (2) the equilibria corresponding to the memory vectors have to be asymptotically stable. Two categories of associative

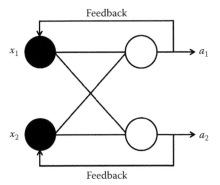

FIGURE 2.7 A generalized recurrent neural network (RNN). The output of one layer can route back (feedback) to the input layer of a previous layer. Feedback is modified by a set of weights as to enable automatic adaptation through a specified learning process.

memories are encountered: (1) autoassociative and (2) heteroassociative. Kohonen (1984) characterized autoassociative memory in terms of its operation, whereby an incomplete key pattern is replenished into a complete (stored) version. Because of this characteristic, autoassociative memories are extensively used in pattern recognition applications. For heteroassociative memory, an output pattern y_k is selectively produced in response to an input pattern x_k, with the paired associates x_k and y_k freely selected and independent of each other (Kohonen, 1984).

Dynamic systems are composed of two unique components: (1) the state (multivariate vector of variables, parameterized with respect to time) and (2) the dynamic (describes how the state evolves through time). The dynamics of RNNs have been routinely studied, including discussions on the differences between discrete systems (evolving in discrete time) and those exhibiting continuous characteristics (evolving in continuous time) (e.g., Markarov et al., 2008). Studies have provided both qualitative insight and, in many cases, quantitative formulas for predicting the dynamical behavior of particular systems and how that behavior changes as network parameters are varied (e.g., Beer, 1995; Parlos et al., 2000; Yu, 2004). A common assumption among many of these studies investigating the dynamical analysis of continuous-time neural network systems is that the activation functions are differentiable and bounded, such as the standard sigmoid-type functions employed in most conventional neural networks.

Depending on the density of the feedback connections present, RNNs can take on a simple or partial recurrent architecture (e.g., Elman model, Jordan model) or a fully or total recurrent architecture (Hopfield model). In simple RNNs, partial recurrence is created by feeding back delayed hidden unit outputs or the outputs of the network as additional model inputs (Elman, 1990). In such networks, groups are updated in the order in which they appear in the network's group array. A group update in this case consists of computing its inputs and immediately computing its outputs. Total recurrent networks use fully interconnected architectures and learning algorithms that can deal with time-varying input and/or output in a rather complex fashion. In contrast to simple recurrent networks, all groups update their inputs and then have all groups compute their outputs simultaneously. A unique property of recurrent-type networks is that their state can be described by an energy function. The energy function is used to prove the stability of recurrent-type networks, where it assumes locally minimal values at stable states. The energy function can also be generalized to arbitrary vectors **x** and **y**. The next section deals specifically with the Hopfield model, which will extend our discussion on the energy function and recurrent neural network characteristics.

2.3.1 THE HOPFIELD NETWORK

Hopfield's seminal paper in 1982 led to a greater understanding and acceptance of autoassociative models (Hopfield, 1982). In autoassociative models, the inputs and outputs map to the same state. If the neural network recognizes a pattern, it will return that pattern instinctively. As a learning algorithm for storing such patterns, the incorporation of Hebb's rule (Hebb, 1949) concepts is well documented. Hebbian learning will be discussed in detail in Chapter 3. Hopfield networks are used as an autoassociative memory with fully recurrent connections between the input and output (Figure 2.8). The Hopfield network consists of a single layer of processing elements where each neuron is connected to every other neuron in the network other than itself through weighted synaptic links, w_{ij}. The connection weight matrix **W** is

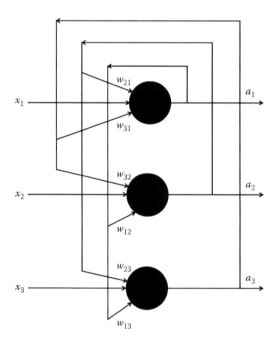

FIGURE 2.8 A characteristic Hopfield network with fully recurrent connections between the input and output. Notice that there is no self-feedback in this network. Neurons present are connected to all other neurons except themselves, with the resulting weight matrix being symmetric.

square and symmetric, that is, $w_{ij} = w_{ji}$ for $i, j = 1, 2, ..., n$ (Hopfield, 1982). Note that each unit has an additional external input I_i. This external input leads to a modification in the computational approach (Hopfield, 1982):

$$\text{input}_j = \sum_{i=1}^{n} x_i w_{ij} + I_j \tag{2.7}$$

for $j = 1, 2, ..., n$. A single associated pattern pair is stored by computing the weight matrix as follows:

$$\mathbf{W}_k = \mathbf{X}_k{}^T \mathbf{Y}_k \tag{2.8}$$

where $\mathbf{Y}_k = \mathbf{X}_k$, and:

$$W = \alpha \sum_{k-1}^{p} W_k \tag{2.9}$$

to store p different associated pattern pairs. Recall that the Hopfield model is an autoassociative memory model, and therefore patterns rather than associated pattern pairs are stored in memory. The output of each neuron is in feedback formation, via a delay element, to each of the other neurons in the model, and given by

$$\text{output} = f\left(\sum_{j \neq i} W_{ij} \text{ input}_j \right) \tag{2.10}$$

Hopfield networks can be classified into discrete and continuous forms based on their output functions. For both, Hopfield (1982) introduced energy functions to demonstrate model stability and to specify the weight matrix of the network. In discrete Hopfield networks, units employ a bipolar output function where the states of the units are zero; that is, the output of the units remains the same if the current state is equal to some threshold value (Wen et al., 2009). Hopfield (1982) developed the following energy function to demonstrate discrete model stability:

$$E_{\text{discrete}} = -\frac{1}{2} \sum_{i=1}^{n} \sum_{j=1}^{n} x_i w_{ij} x_j - \sum_{i=1}^{n} x_i I_i \tag{2.11a}$$

The foregoing energy notation hosts an underlying Lyapunov function for the activity dynamics. A Lyapunov function is a function of a vector and of time that is used to test whether a dynamical system is stable. Starting in any initial state, the state of the system evolves to a final state that is a (local) minimum of the Lyapunov function (Hopfield, 1982).

The continuous Hopfield network is a generalization of the discrete case, with common output functions exploited being sigmoid and hyperbolic tangent functions. Hopfield (1984) proposed the following energy function to demonstrate continuous model stability:

$$E_{\text{continuous}} = -\frac{1}{2}\sum_{i=1}^{n}\sum_{j=1}^{n} x_i w_{ij} x_j - \sum_{i=1}^{n} x_i I_i + \sum_{i=1}^{n}\left(\frac{1}{R_i}\right)\int_{0}^{x_i} g_i^{-1}(x_i)d \quad (2.11b)$$

where the function $g_i^{-1}(x_i)$ is a monotone increasing function: a function whose value increases when that of a given variable increases, and decreases when the latter is subsequently diminished. The continuous model has been particularly useful in optimization applications, where it is considered superior to the discrete case in terms of the local minimum problem as a result of its smoother energy surface (Šíma and Orponen, 2003; Wen et al., 2009).

An additional neural network used for optimization, the Boltzmann machine, is an extension of discrete Hopfield networks. The goal of Boltzmann learning is to produce a neural network that categorizes input patterns according to a Boltzmann distribution. In mathematics, it is a recognized distribution function or probability measure for the distribution of the states of a system under investigation. Similar to Hopfield networks, Boltzmann machines have symmetric connections between neurons. Boltzmann machines are fundamentally an extension of simple stochastic associative networks to include hidden neurons (units not involved in the pattern vector) (Ackley et al., 1985). The displayed hidden neurons allow such networks to model data distributions that are much more complex than those that can be learned by Hopfield networks. In addition, Boltzmann machines use stochastic neurons with probabilistic firing mechanisms (Hilton and Sejnowski, 1968). Standard Hopfield networks use neurons based on the McCulloch–Pitts model with a deterministic firing.

Although not extensively applied in biological and environmental analyses, the Boltzmann machine learning offers great potential, especially in regards to agent-based simulation modeling (Bass and Nixon, 2008). Variations of the original Boltzmann machine developed by Sejnowski and colleagues (Ackley et al., 1985) include higher-order, conditional, restricted, and mean field machines, with each varying slightly in its approach to learning rules. An assortment of informative papers is available for more extensive examination of both theoretical concepts of Boltzmann machines and related applications (Kappen, 1995; Leisink and Kappen, 2000; Bass and Nixon, 2008).

2.3.2 KOHONEN'S SELF-ORGANIZING MAP

Many natural systems exhibit self-organization processes (typically associated with nonlinear phenomena) where the development of new and complex structures can take place within the system without considerable influence from the environment.

All the intricacies associated with nonlinearity can be understood through the interplay of positive and negative feedback cycles. For example, the spontaneous folding of proteins and other biomacromolecules would constitute the contribution of self-organization in biological systems. Studies have also shown that bacterial colonies display complex collective dynamics, frequently culminating in the formation of biofilms and other order superstructure (e.g., Cho et al., 2007). This self-organization may be central to early-stage organization of high-density bacterial colonies populating small, physically confined growth niches, including those implicated in infectious diseases.

Complex nonlinear dynamic systems are also ubiquitous in the landscapes and phenomena studied by earth scientists. Such concepts as chaos, fractals, and self-organization, originating from research in nonlinear dynamics, have proved to be powerful approaches to understanding and modeling the evolution and characteristics of a wide variety of landscapes and bedforms (Baas, 2002). Additionally, soil formation, in the broadest sense, is the result of the synergetic processes of self-organization of an in situ soil system during its functioning in time and space. In neural computing terms, self-organization is the modification of synaptic strengths that, in turn, underlies the dynamics and behavioral modification of pattern recognition of higher-level representations (Allinson et al., 2002). Unsupervised training, described in detail in Chapter 3, is a process by which networks learn to form their own classifications of the training data without external guidance. For this to occur, class membership must be broadly defined by the input patterns sharing common features. The network must be able to identify those features across a broad range of input patterns.

Relevant to these concepts would be the origin of Kohonen's self-organizing map (SOM) modeled after the self-organization of the neural links between the visual cortex and the retina cells when excited by independent stimuli (Kohonen, 1982). Kohonen's SOM can be viewed as a model of unsupervised learning, and as an adaptive knowledge representation scheme. This SOM displays a feedforward structure with a single computational layer arranged in rows and columns (Figure 2.9). As shown, each neuron is effusively connected to all the sources in the input layer. Consider the output cells in the computational layer as inhabiting positions in space and arranged in a grid pattern, so that it makes sense to talk about those output cells that are neighbors of a particular cell. Each neuron in the output layer is represented by a D-dimensional weight vector $\mathbf{x} = \{x_i : i = 1, \ldots, D\}$, where D is equal to the dimension of the input vectors. The neurons are connected to adjacent neurons by a neighborhood relation that influences the topology or structure of the SOM map, with the connection weights between the input units i and the neurons j in the computational layer written as $\mathbf{w}j = \{w_{ji} : j = 1,\ldots,N; i = 1,\ldots, D\}$. Here, N = the total number of neurons.

The SOM combines two paradigms of unsupervised learning: (1) clustering (grouping data by similarity) and (2) projection methods (extraction of explanatory variables). The Kohonen computational layer processing elements each measure the Euclidean distance of its weights from the incoming input values. The Euclidean metric is defined as the function

$$d : \Re^n \times \Re^n \longrightarrow \Re$$

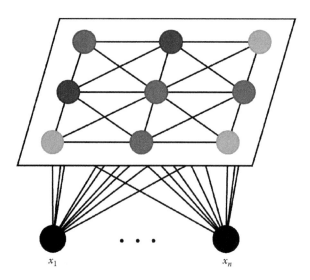

x_1 • • • x_n

FIGURE 2.9 *A color version of this figure follows page 106*. A Kohonen's self-organizing map (SOM) displaying a feedforward structure with a single computational layer arranged in rows and columns. For the input layer, each node is a vector representing N terms. Each output node is a vector of N weights. Upon visual inspection, colors are clustered into well-defined regions, with regions of similar properties typically found adjoining each other. (From Hanrahan, G. 2010. *Analytical Chemistry*, 82: 4307–4313. With permission from the American Chemical Society.)

that assigns to any two vectors in Euclidean n-space $\mathbf{x} = (x_1, \ldots, x_n)$ and $y = (y_1, \ldots, y_n)$:

$$d(x,y) = \sqrt{(x_1 - y_1)^2 + \ldots (x_n - y_n)^2} \qquad (2.12)$$

The Kohonen element with the minimum distance (i.e., closest to the input vector) is the winner and outputs a one to the output layer (Kohonen, 1996). The winning processing element thus represents the input value in the Kohonen two-dimensional map. The input data are represented by a two-dimensional vector that preserves the order of the higher-dimensional input data. During training, the Kohonen processing element with the smallest distance adjusts its weight to be closer to the values of the input data. The neighbors of the winning element also adjust their weights to be closer to the same input data vector. The adjustment of neighboring processing elements is influential in preserving the order of the input space. Training is done with the competitive Kohonen learning law, which is described in detail in Chapter 3.

The SOM introduced by Kohonen has provided a definitive tool for numerous applications, including data mining, classification, analysis, and visualization. Although introduced by Kohonen as a heuristic, SOMs have been well received as statistical methods of late, which has led to more firm theoretical acceptance (Hassenjäger et al., 1999). SOMs map nonlinear relationships between high-dimensional input data into simple geometric relationships on a two-dimensional grid. The mapping attempts to preserve the most important topological and metric relationships of the original data elements and, thus, inherently clusters the data (Kohonen, 1996). As stated by Furukawa (2009), Kohonen's SOM is one of the best techniques currently available for visualizing the distribution of each class in an entire labeled data set. Kohonen (1995) showed that a combination of the basic SOM method with supervised learning allowed the extension of the scope of applications to labeled data. Hammer and colleagues (Hammer et al., 2004), in a detailed review, highlighted a variety of extensions of SOMs and alternative unsupervised learning models that exist that differ from the standard SOM with respect to neural topology or underlying dynamic equations. At last, its use is becoming progressively more evident in biological and environmental analyses, where efficient data visualization and clustering is needed to deal with complex data sets. Used unaccompanied, or in combination with other techniques (e.g., principal component analysis [PCA]), Kohonen's SOM is capable of, for example, clustering ecosystems in terms of environmental conditions. Compared with traditional clustering and ranking approaches, this combination is reported to have considerable advantages (e.g., Tran et al., 2003), including the capacity to impart an effective means of visualizing complex multidimensional environmental data at multiple scales and offering a single ranking needed for environmental assessment while still enabling more detailed analyses.

2.4 CONCLUDING REMARKS

The objective of this chapter is to provide readers with a lucid and detailed survey of fundamental neural network architecture principles, where neuron connectivity and layer arrangement dictate advanced understanding and applicability. It was shown that combinations of interconnected computing units can act as simple models for complex neurons in biological systems with the arrangement of layers and neurons, nodal connectivity, and nodal transfer functions being of paramount importance in neural network performance. Discussion evolved from simple feedforward structures, those that carry information with no directed cycle, to recurrent structures with bidirectional data flow demonstrating partial recurrent or total recurrent architectures. Simple examples that demonstrate the effect of interconnected neuronal structures were provided, that is, structure that is manifest at the level of the practitioner. Although not fully comprehensive, this chapter has accounted for the beginning stages of neural network development, and the commitment to applying models with good performance and applicability required for routine and experimental study of complex natural systems.

REFERENCES

Ackley, D., Hinton, G., and Sejnowski, T. 1985. Learning algorithm for Boltzmann machines. *Cognitive Science* 9: 147–169.
Adeli, H., and Panakkat, A. 2009. Probabilistic neural network for earthquake magnitude prediction. *Neural Networks* doi:10.1016/j.neunet.2009.05.003.
Agatonovic-Kustrin, S., and Beresford, R. 2000. Basic concepts of artificial neural network (ANN) modeling and its application in pharmaceutical research. *Journal of Pharmaceutical and Biomedical Analysis* 22: 717–727.

Allinson, N., Yin, H., and Obermayer, K. 2002. Introduction: New developments in self-organizing maps. *Neural Networks* 15: 943.

Andrieu, C., de Freitas, N., and Dorcet, A. 2001. Robust full Bayesian learning for radial basis networks. *Neural Computation* 13: 2359–2407.

Baas, A.C.W. 2002. Chaos, fractals and self-organization in coastal geomorphology: Simulating dune landscapes in vegetated environments. *Geomorphology* 48: 309–328.

Bass, B., and Nixon, T. 2008. Boltzmann learning. In: *Encyclopedia of Ecology*. Elsevier: Amsterdam, pp. 489–493.

Beer, R. 1995. On the dynamics of small continuous-time recurrent neural networks. *Adaptive Behavior* 3: 469–509.

Bishop, C. 2005. *Neural Networks for Pattern Recognition*. Oxford University Press: Oxford.

Botvinick, M.M., and Plaut, D. 2006. Short-term memory for serial order: A recurrent neural network model. *Psychological Review* 113: 201–233.

Bugmann, G. 1998. Classification using networks of normalized radial basis functions. *International Conference on Advances in Pattern Recognition*. Plymouth, U.K., November 23–25, 1998.

Chen, S., Cowan, C.F.N., and Grant. 1991. Orthogonal least squares learning algorithm for radial basis function networks. *IEEE Transactions on Neural Networks* 2: 302–309.

Cho, H., Jönsson, H., Campbell, K., Melke, P., Williams, J.W., Jedynak, B., Stevens, A.M., Groisman, Z., and Levchenko, A. 2007. Self-organization in high-density bacterial colonies: Efficient crowd control. *PLoS Biology* 5: 2614–2623.

Daqi, G., and Yan, J. 2005. Classification methodologies of multilayer perceptrons with sigmoid activation functions. *Pattern Recognition* 38: 1469–1482.

Duch, W., Adamczak, R., and Diercksen, G.H.F. 1999. Distance-based multilayer perceptrons. In M. Mohammadian (Ed.) *Computational Intelligence for Modelling, Control and Automation*. IOS Press: Amsterdam.

Elman, J.L. 1990. Finding structure in time. *Cognitive Science* 14: 179–211

Fausett, L. 1994. *Fundamentals of Neural Networks: Architecture, Algorithms, and Applications*. Prentice Hall: Upper Saddle River, NJ.

Feng, L., and Hong, W. 2009. Classification error of multilayer perceptron neural networks. *Neural Computing and Applications* 18: 377–380.

Furukawa, T. 2009. SOM of SOMs. *Neural Networks* 22: 463–478.

García-Altés, A. 2007. Applying artificial neural networks to the diagnosis of organic dyspepsia. *Statistical Methods in Medical Research* 16: 331–346.

Gardner, M.W., and Dorling, S.R. 1998. Artificial neural networks (the multilayer perceptron): A review of applications in the atmospheric sciences. *Atmospheric Environment* 32: 2627–2636.

Gelman, A., Carlin, J., Stern, H., and Rubin, D. 2003. *Bayesian Data Analysis*. CRC Press: Boca Raton, FL.

Hagan, M.T., Demuth, H.B., and Beale, M.H. 1996. *Neural Network Design*. PWS Publishing Company: Boston.

Hammer, B., Micheli, A., Sperduti, A., and Strickert, M. 2004. Recursive self-organizing network models. *Neural Networks* 17: 1061–1085.

Hanrahan, G. 2010. Computational neural networks driving complex analytical problem solving. *Analytical Chemistry* 82: 4307–4313.

Hassenjäger, M., Ritter, H., and Obermayer, K. 1999. Active learning in self-organizing maps. In E. Oja and S. Kaski (Eds.), *Kohonen Maps*. Elsevier: Amsterdam.

Hassoun, M. H. 1993. *Associative Neural Memories: Theory and Implementation*. Oxford Press: Oxford.

Hebb, D.O. 1949. *The Organization of Behaviour*. John Wiley & Sons: New York.

Hilton, G.E., and Sejnowski, T.J. 1968. Learning and relearning in Boltzmann machines. In D.E. Rumelhart (Ed.), *Parallel Distributed Processing: Explorations in the Microstructure of Cognition.* MIT Press: Cambridge, MA, pp. 282–317.

Hopfield, J.J. 1982. Neural networks and physical systems with emergent collective computational properties. *Proceedings of the National Academy of Sciences* 79: 2554–2558.

Hopfield, J.J. 1984. Neurons with graded response have collective computational properties like those of two-state neurons. *Proceedings of the National Academy of Sciences* 81: 3088–3092.

Jalali-Heravi, M. 2008. Neural networks in analytical chemistry. In D.S. Livingstone (Ed.), *Artificial Neural Networks: Methods and Protocols.* Humana Press: NJ.

Kappen, H.J. 1995. Deterministic learning rules for Boltzmann machines. *Neural Networks* 8: 537–548.

Kohonen, T. 1982. Self-organized formation of topologically correct feature maps. *Biological Cybernetics* 43: 59–69.

Kohonen, T. 1984. *Self-Organization and Associative Memory.* Springer-Verlag: Berlin.

Kohonen, T. 1995. Learning vector quantization. In M. Aribib (Ed.), *The Handbook of Brain Theory and Neural Networks.* MIT Press: Cambridge, MA., pp. 537–540.

Kohonen, T. 1996. *Self-Organization and Associative Memory.* 3rd edition, Springer-Verlag: Berlin.

Lampinen, J., and Vehtar, A. 2001. Bayesian approach for neural networks-review and case studies. *Neural Networks* 14: 257–274.

Leisink, M.A.R., and Kappen, H.J. 2000. Learning in higher order Boltzmann machines using linear response. *Neural Networks* 13: 329–335.

Markarov, V.A., Song, Y., Velarde, M.G., Hübner, D., and Cruse, H. 2008. Elements for a general memory structure: Properties of recurrent neural networks used to form situation models. *Biological Cybernetics* 98: 371–395.

May, R.J., Maier, H.R., and Dandy, G.C. 2009. Developing artificial neural networks for water quality modelling and analysis. In G. Hanrahan (Ed.), *Modelling of Pollutants in Complex Environmental Systems.* ILM Publications: St. Albans, U.K.

Mielniczuk, J., and Tyrcha, J. 1993. Consistency of multilayer perceptron regression estimators. *Neural Networks* 6: 1019–1022.

Minsky, M.L., and Papert, S.A. 1969. *Perceptrons: An Introduction to Computational Geometry.* MIT Press: Cambridge, MA.

Parlos, A.G., Rais, O.T., and Atiya, A.F. 2000. Multi-step-ahead prediction using dynamic recurrent neural networks. *Neural Networks* 13: 765–786.

Priddy, K.L., and Keller, P.E. 2005. *Artificial Neural Networks: An Introduction.* SPIE Press: Bellingham, WA.

Schiøler, H., and Hartmann, U. 1992. Mapping neural network derived from the Parzen window estimator. *Neural Networks* 5: 903–909.

Seidel, P., Seidel, A., and Herbarth, O. 2007. Multilayer perceptron tumour diagnosis based on chromatography analysis of urinary nucleosides. *Neural Networks* 20: 646–651.

Šíma, J., and Orponen, P. 2003. Continuous-time symmetric Hopfield nets are computationally universal. *Neural Computation* 15:693–733.

Specht, D.F. 1990. Probabilistic neural networks. *Neural Networks* 3: 110–118.

Specht, D.F. 1991. A generalized regression neural network. *IEEE Transactions on Neural Networks* 2: 568–576.

Tomandl, D., and Schober, A. 2001. A modified general regression neural network (MGRNN) with new, efficient algorithms as a robust "black box"-tool for data analysis. *Neural Networks* 14: 1023–1034.

Tran, L.T., Knight, C.G., O'Niell, R.V., Smith, E.R., and O'Connell, M. 2003. Self-organizing maps for integrated environmental assessment of the Mid-Atlantic region. *Environmental Management* 31: 822–835.

Wen, U-P., Lan, K-M., and Shih, H-S. 2009. A review of Hopfield neural networks for solving mathematical programming problems. *European Journal of Operational Research* 198: 675–687.

Widrow, B., and Lehr, M.A. 1990. 30 years of adaptive neural networks: Perceptron, Madeline, and backpropagation. *Proceeding of the IEEE* 78, 1415–1442.

Yu, W. 2004. Nonlinear system identification using discrete-time recurrent neural networks with stable learning algorithms. *Information Sciences* 158: 131–147.

Yuste, A.J., and Dorado, M.P. 2006. A neural network approach to simulate biodiesel production from waste olive oil. *Energy Fuels* 20: 399–402.

Zhou, J. Han, Y., and So, S-S. 2008. Overview of artificial neural networks. In D.S. Livingstone (Ed.), *Artificial Neural Networks: Methods and Protocols*. Humana Press: NJ.

Zhu, B. 2006. Voronoi diagram and Delaunay triangulation: Applications and challenges in bioinformatics. *Proceedings of 3rd International Symposium on Voronoi Diagrams in Science and Engineering 2006,* Number 4124794, pp. 2–3.

Zomer, S. 2004. Classification with Support Vector Machines. Homepage of Chemometrics, Editorial: http://www.chemometrics.se/images/stories/pdf/nov2004.pdf.

3 Model Design and Selection Considerations

3.1   IN SEARCH OF THE APPROPRIATE MODEL

The fact that neural network models can approximate virtually any measurable function up to an arbitrary degree of accuracy has been emphasized. Moreover, knowledge of neuron connectivity and the choice of layer arrangement have been equally covered. With this understanding comes the daunting task of selecting the most appropriate model for a given application. This is especially true in the study of natural systems, given the unpredictable availability, quality, representativeness, and size of input data sets. Although no perfect model exists for a given application, modelers have the ability to develop and implement effective neural network techniques based on a coordinated framework such as that depicted in Figure 3.1. This circumvents many of the mistakes modelers make in their experimental approach: a line of attack with poorly stated purposes and no clear indication of model performance expectancy or validation criteria.

The need for consistency in the way we model natural systems is evident, particularly when considering the absence of methodological protocols when scouring the literature (e.g., as detailed by Arhonditsis and Brett, 2004). Modelers must also be cognizant of structured methods for adequately assessing the uncertainty underlying model predictions and simulation processes when modeling natural systems. Arhonditsis (2009) states:

> Model input error stems mainly from the uncertainty underlying the values of model parameters, initial conditions and forcing functions, as well as from the realization that all models are drastic simplifications of reality that approximate the actual process: that is, essentially, all parameters are effective values (e.g., spatially and temporally averaged), unlikely to be represented by fixed constraints.

If we think in terms of neural networks, such models can be developed and trained to approximate the functional relationships between the input variables and, for example, the uncertainty descriptors estimated from Monte Carlo (MC) realizations in hydrological modeling (Shrestha et al., 2009), thus deriving meaningful uncertainty bounds of associated model simulations. This is one example among many that demonstrate the use of neural network models in assessing uncertainty, and therefore aiding in the development of effective models for the study of natural systems. As a guide to those incorporating neural network models in their research efforts, this chapter details an eight-step process and presents fundamental building blocks of

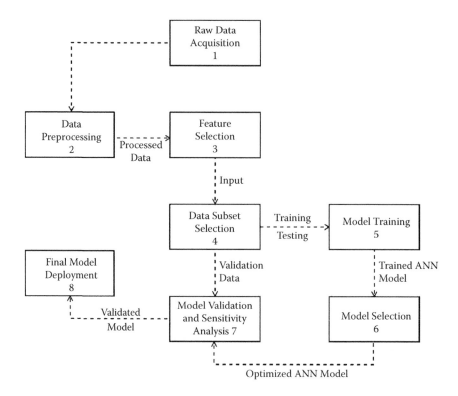

FIGURE 3.1 Model framework for neural network development and final application. (Based on an original figure from Hanrahan, G. 2010. *Analytical Chemistry*, 82: 4307–4313. With permission from the American Chemical Society.)

model design, selection, validation, sensitivity and uncertainty analysis, and application from a statistically sound perspective.

3.2 DATA ACQUISITION

In biological and environmental analysis activities, it is generally assumed that the objectives of such activities have been well considered, the experimental designs have been structured, and the data obtained have been of high quality. But as expected, biological and environmental systems are inherently complex and data sets generated from modern analytical methods are often colossal. For example, high-throughput biological screening methods are generating expansive streams of information about multiple aspects of cellular activity. As more and more categories of data sets are created, there is a corresponding need for a multitude of ways in which to correctly and efficiently data-mine this information.

Complex environmental and biological systems can also be explicitly linked. For example, the analysis of the dynamics of specific microbial populations in the context of environmental data is essential to understanding the relationships between microbial community structure and ecosystem processes. Often, the fine-scale phylogenetic resolution acquired comes at the expense of reducing sample through-put and sacrificing complete sampling coverage of the microbial community (Jacob et al., 2005). The natural variability in environmental data (e.g., seasonality, diurnal trends, and scale) typically accounts for incomplete or nonrepresentative sampling and data acquisition. Being able to capture such variability ensures that the neural network input data are appropriate for optimal training to effectively model environmental systems behavior. Moreover, it is essential to develop a sampling protocol considering the identification of suitable scientific objectives, safety issues, and budget constraints.

3.3 DATA PREPROCESSING AND TRANSFORMATION PROCESSES

Commonly used as a preliminary data mining practice, data preprocessing transforms the data into a format that will be more easily and effectively processed for the purpose of the practitioner (e.g., in neural network applications). This is especially important in biological and environmental data sets, where missing values, redundancy, nonnormal distributions, and inherently noisy data are common and not the exception. One of the most common mistakes when dealing with raw data is to simply dispense with observations regarding missing variable values, even if it is only one of the independent variables that is missing. When considering neural networks, learning algorithms are affected by missing data since they rely heavily on these data to learn the underlying input/output relationships of the systems under examination (Gill et al., 2007). The ultimate goal of data preprocessing is thus to manipulate the unprocessed data into a form with which a neural network can be sufficiently trained. The choices made in this period of development are crucial to the performance and implementation of a neural network. Common techniques used are described in detail in the following text.

3.3.1 HANDLING MISSING VALUES AND OUTLIERS

Problems of missing data and possible outliers are inherent in empirical biological and environmental science research, but how these are to be handled is not adequately addressed by investigators. There are numerous reasons why data are missing, including instrument malfunction and incorrect reporting. To know how to handle missing data, it is advantageous to know why they are absent. Practitioners typically consider three general "missingness mechanisms," where missing data can be classified by their pattern: missing completely at random (MCAR), missing at random (MAR), or missing not at random (MNAR) (Sanders et al., 2006). Inappropriate handling of missing data values will alter analysis because, until demonstrated otherwise, the practitioner must presuppose that missing cases differ in analytically significant ways from cases where values are present.

Investigators have numerous options with which to handle missing values. An all too often-used approach is simply ignoring missing data, which leads to limited efficiency and biases during the modeling procedure. A second approach is data imputation, which ranges from simple (e.g., replacing missing values by zeros, by the row average, or by the row median) to more complex multiple imputation mechanisms (e.g., Markov chain Monte Carlo [MCMC] method). An alternative approach developed by Pelckmans et al. (2005) made no attempt at reconstructing missing values, but rather assessed the impact of the missingness on the outcome. Here, the authors incorporated the uncertainty due to the missingness into an appropriate risk function. Work by Pantanowitz and Marwala (2009) presented a comprehensive comparison of diverse paradigms used for missing data imputation. Using a selected HIV seroprevalence data set, data imputation was performed through five methods: Random Forests, autoassociative neural networks with genetic algorithms, autoassociative neuro-fuzzy configurations, and two random forest and neural-network-based hybrids. Results revealed that Random Forests was far superior in imputing missing data for the given data set in terms of accuracy and in terms of computation time. Ultimately, there are numerous types of missing-data-handling methods available given the different types of data encountered (e.g., categorical, continuous, discrete), with no one method being universally suitable for all applications.

Unlike the analysis of missing data, the process of outlier detection aims to locate "abnormal" data records that are considered isolated from the bulk of the data gathered. There are a number of probable ways in which to ascertain if one or more values are outliers in a representative data set. If the values are normally distributed, then an investigator can isolate outliers using statistical procedures (e.g., Grubbs' test, Dixon's test, stem and leaf displays, histograms, and box plots). If the values have an unidentified or nonstandard distribution, then there exist no prevailing statistical procedures for identifying outliers. Consider the use of the k-NN method, which requires calculation of the distance between each record and all other records in the data set to identify the k-NN for each record (Hodge et al., 2004). The distances can then be examined to locate records that are the most distant from their neighbors and, hence, values that may correspond to outliers. The k-NN approach can also be used with missing data by exchanging missing values with the closest possible available data using the least distance measure as matching criteria.

3.3.2 LINEAR SCALING

Linear scaling is a somewhat simple technique used in normalizing a range of numeric values with the linear scaling transformation given by

$$z_{ij} = \frac{x_{ij} - \min(x_j)}{\max(x_j) - \min(x_j)} \qquad (3.1)$$

where z_{ij} = a single element in the transformed matrix $\mathbf{Z}$. Here, the original data matrix $\mathbf{X}$ is converted to $\mathbf{Z}$, a new matrix in which each transformed variable has both a minimum and maximum value of unity. This approach is particularly useful in neural network algorithms, given the fact that many assume a specified input variable range to avoid saturation of transfer functions in the hidden layer of the MLP

(May et al., 2009). At a bare minimum, data must be scaled into the range used by the input neurons in the neural network. This is characteristically in the range of −1 to 1 or zero to 1. The input range compulsory for the network must also be established. This means of normalization will scale the input data into a suitable range but will not increase their uniformity.

3.3.3 AUTOSCALING

Studies have also shown the usefulness of the autoscaling transformation in enhancing the classification performance of neural networks by improving both the efficiency and the effectiveness of associated classifiers. Assessing the performance of the autoscaling transformation can be accomplished by estimating the classification rate of the overall pattern recognition process. Autoscaling involves mean-centering of the data and a division by the standard deviation of all responses of a particular input variable, resulting in a mean of zero and a unit standard deviation of each variable:

$$z_{ij} = \frac{x_{ij} - \overline{x}_j}{\sigma_{xj}}$$

(3.2)

where x_{ij} = response of the ith sample at the jth variable, $\overline{x}_j$ = the mean of the jth variable, and σ_{xj} = the standard deviation of the jth variable.

3.3.4 LOGARITHMIC SCALING

The logarithmic transformation is routinely used with data when the underlying distribution of values is nonnormal, but the data processing technique assumes normality (Tranter, 2000). This transformation can be advantageous as by definition, log-normal distributions are symmetrical again at the log level. Other transformation techniques used in skewed distributions include inverse and square root transformations, each having distinct advantages depending on the application encountered.

3.3.5 PRINCIPAL COMPONENT ANALYSIS

Often, data sets include model input variables that are large in scope, redundant, or noisy, which can ultimately hide meaningful variables necessary for efficient and optimized modeling. In such cases, methods based on principal component analysis (PCA) for preprocessing data used as input neural networks are advantageous. PCA reduces the number of observed variables to a smaller number of principal components (PCs). Each PC is calculated by taking a linear combination of an eigenvector of the correlation matrix with an original variable. As a result, the PCA method determines the significance of the eigenvalues of the correlation matrix associated with the first PCs of the data in order to select the subset of PCs for the sample that provides the optimum generalization value. In terms of neural network modeling, the PCs with larger eigenvalues represent the more relative amount of variability of the training data set. The largest PC can be first applied as an input variable of a corresponding MLP, with subsequent PCs employed as MLP input data sets. The process continues until all the PCs that represent the majority of the variability of the training data set are included in the input data set of the corresponding MLP.

PCA input selection has been shown to be useful in classifying and simplifying the representation of patterns and improving the precision of pattern recognition analysis in large (and often redundant) biological and environmental data sets. For example, Dorn et al. (2003) employed PCA input selection to biosignature detection. Results showed that PCA correctly identified glycine and alanine as the amino acids contributing the most information to the task of discriminating biotic and abiotic samples. Samani et al. (2007) applied PCA on a simplified neural network model developed for the determination of nonleaky confined aquifer parameters. This method avoided the problems of selecting an appropriate trained range, determined the aquifer parameter values more accurately, and produced a simpler neural network structure that required less training time compared to earlier approaches.

3.3.6   WAVELET TRANSFORM PREPROCESSING

The wavelet transform (WT) has been shown to be a proficient method for data compression, rapid computation, and noise reduction (Bruce et al., 1996). Moreover, wavelet transforms have advantages over traditional methods such as Fourier transform (FT) for representing functions that have discontinuities. A wavelet is defined as a family of mathematical functions derived from a basic function (wavelet basis function) by dilation and translation (Chau et al., 2004). In the interest of brevity, a detailed theoretical explanation will be left to more specialized sources (e.g., Chau et al., 2004; Cooper and Cowan, 2008). In general terms, the continuous wavelet transform (CWT) can be defined mathematically as (Kumar and Foufoula-Georgiou, 1997)

$$W_f(\lambda, t) = \int_{-\infty}^{\infty} f(u)\overline{\Psi}_{\lambda, t}(u)\, du \ \ \lambda > 0, \tag{3.3}$$

where

$$\Psi_{\lambda, t}(u) \equiv \frac{1}{\sqrt{\lambda}}\Psi\left(\frac{u - t}{\lambda}\right) \tag{3.4}$$

represents the wavelets (family of functions), λ = a scale parameter, t = a location parameter, and $\overline{\Psi}_{\lambda, t(u)}$ = the complex conjugate of $\Psi_{\lambda, t(u)}$. The inverse can be defined as

$$f(t) = \frac{1}{C_\psi} \int_{-\infty}^{\infty} \int_{0}^{\infty} \lambda^{-2} Wf(\lambda, u)\psi_{\lambda, u}(t)\, d\lambda\, du \tag{3.5}$$

where C_ψ = a constant that depends on the choice of wavelet. Wavelet transforms implemented on discrete values of scale and location are termed discrete wavelet transforms (DWTS). Cai and Harrington (1999) detailed two distinct advantages of utilizing WT preprocessing in the development of neural networks: (1) data compression and (2) noise reduction. The diminution of noise is essential in multivariate analysis because many methods overfit the data if care is not judiciously practiced. The authors also expressed how wavelet compression can intensify the training rate of a neural network, thus permitting neural network models to be built from data that otherwise would be prohibitively large.

3.4 FEATURE SELECTION

Feature selection is one of the most important and readily studied issues in the fields of system modeling, data mining, and classification. It is particularly constructive in numerical systems such as neural networks, where data are symbolized as vectors in a subspace whose components (features) likely correspond to measurements executed on physical systems or to information assembled from the observation of phenomena (Leray and Gallinari, 1999). Given the complexity of biological and environmental data sets, suitable input feature selection is required to warrant robust frameworks for neural network development. For example, MLPs are ideally positioned to be trained to perform innumerable classification tasks, where the MLP classifier performs a mapping from an input (feature) space to an output space. For many classification applications in the biological and environmental sciences, a large number of useful features can be identified as input to the MLP. The defined goal of feature selection is thus to appropriately select a subset of k variables, S, from an initial candidate set, C, which comprises the set of all potential model inputs (May et al., 2008). Optimal subset selection results in dimension reduction, easier and more efficient training, better estimates in the case of undersized data sets, more highly developed processing methods, and better performance.

Algorithms for feature selection can be characterized in terms of their evaluation functions. In the feature wrapper approach, outlined in Figure 3.2, an induction algorithm is run on partitioned training and test data sets with different sets of features removed from the data. It is the feature subset with optimal evaluation that is selected and used as the final data set on which induction learning is carried out. The resulting classifier is then put through a final evaluation using a fresh test data set independent of the primary search. Wrapper implementation can be achieved by forward selection, backward elimination, or global optimization (e.g., evolutionary neural networks) (May et al., 2008). These processes are described in detail in a paper by Uncu and Türkşen (2007).

In contrast to wrappers, feature filters use a heuristic evaluation function with features "filtered" independently of the induction algorithm. Feature selection is first based on a statistical appraisal of the level of dependence (e.g., mutual information and Pearson's correlation) between the candidates and output variables (May et al., 2008). The second stage consists of estimating parameters of the regression model on the selected subset. Finally, embedded approaches incorporate the feature selection directly into the learning algorithm (Bailly and Milgram, 2009). They are particularly

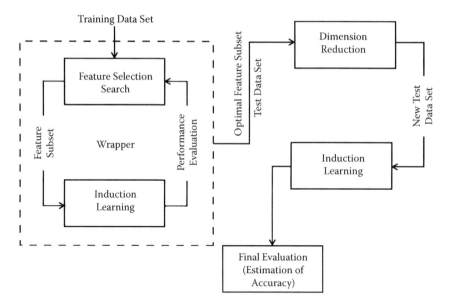

FIGURE 3.2   A schematic of the general feature wrapper framework incorporating induction learning.

useful when the number of potential features is moderately restricted. Novel work by Uncu and Türkşen (2007) presented a new feature selection algorithm that combined feature wrapper and feature filter approaches in order to pinpoint noteworthy input variables in systems with continuous domains. This approach utilized the functional dependency concept, correlation coefficients, and k-NN method to implement the feature filter and associated feature wrappers. Four feature selection methods (all applying the k-NN method) independently selected the significant input variables. An exhaustive search strategy was employed in order to find the most suitable input variable combination with respect to a user-defined performance measure.

3.5   DATA SUBSET SELECTION

Model generalizability is an important aspect of overall neural network development and a complete understanding of this concept is imperative for proper data subset selection. Generalization refers to the ability of model outputs to approximate target values given inputs that are not in the training set. In a practical sense, good generalization is not always achievable and requires satisfactory input information pertaining to the target and a sufficiently large and representative subset (Wolpert, 1996). In order to effectively assess whether one has achieved their goal of generalizability, one must rely on an independent test of the model. As circuitously stated by Özesmi et al. (2006a) in their study on the generalizability of neural network models in ecological applications:

> A model that has not been tested is only a definition of a system. It becomes a scientific pursuit, a hypothesis, when one starts testing it.

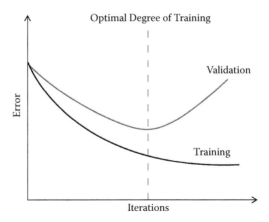

FIGURE 3.3   Profiles for training and validation errors with the optimal degree of training realized at the specified number of iterations where the validation error begins to increase.

To help ensure the possibility of good generalization, modelers split representative data sets into subsets for training, testing, and validation. In neural network modeling practice, it is of the essence to achieve a good balance in the allocation of the input data set, with 90% split between the training (70%) and test sets (20%), and an additional 10% set aside for the validation procedure, routinely reported (e.g., Riveros et al., 2009). The training data set is used for model fitting in computing the gradient and updating the network weights and biases. The validation set is used in modeling assessment where the error on the validation set is supervised during the training phase. Overfitting is an obstacle of fundamental significance during training, with significant implications in the application of neural networks. Ideally, the validation and training errors diminish throughout the initial phase of training. Conversely, when the network begins to overfit the data, the validation error begins to amplify. When the validation error increases for a specified number of iterations, the training is stopped, and the weights and biases at the minimum of the validation error are returned (Figure 3.3). Finally, the test data set is used after the training process in order to formulate a final assessment of the fit model and how acceptably it generalizes. For example, if the error in the test set reaches a minimum at a significantly different iteration value than the validation set error, deficient data partitioning is probable (Priddy and Keller, 2005).

3.5.1   DATA PARTITIONING

Representative and unbiased data subsets have need of sound data partitioning techniques based on statistical sampling methods. Simple random sampling results in all samples of n elements possessing an equal probability of being chosen. Data are indiscriminately partitioned based on a random seed number that follows a standardized distribution between 0 and 1. However, as argued by May et al. (2009), simple random sampling in neural network model development when applied to nonuniform data (such as those found in many biological and environmental applications) can be problematic and result in an elevated degree of variability. Stratified sampling entails a two-step process: (1) randomly partitioning the data into stratified target groups based on a categorical-valued target variable and (2) choosing a simple random sample from within each group. This universally used probability technique is believed to be superior to random sampling because it reduces sampling error. Systematic sampling involves deciding on sample members from a larger population according to an arbitrary starting point and a fixed, periodic interval. Although there is some deliberation among users, systematic sampling is thought of as being random, as long as the periodic interval is resolved beforehand and the starting point is random (Granger and Siklos, 1995).

3.5.2 Dealing with Limited Data

A potential problem in application-based research occurs when available data sets are limited or incomplete. This creates a recognized predicament when developing independent and representative test data sets for appraising a neural network model's performance. As reported by Özesmi et al. (2006b), studies with independent test data sets are uncommon, but an assortment of statistical methods, including K-fold cross-validation, leave-one-out cross-validation, jackknife resampling, and bootstrap resampling, can be integrated. The K-fold cross-validation approach is a two-step process. At the outset, the original data set of m samples is partitioned into K sets (folds) of size m/K. A lone subsample from the K sets is then retained as the validation data for testing the neural model, and the remaining $K - 1$ subsamples are used as dedicated training data. This process is replicated K times, with each of the K subsamples used just once as validation data. The K results from the folds can then be combined to generate a single inference.

The leave-one-out cross-validation method is comparable to K-folding, with the exception of the use of a single sample from the original data set as the validation data. The outstanding samples are used as the training data. This is repeated such that each observation in the sample is used only once as the validation data. As a result, it is time and again considered computationally expensive because of the large number of times the training process is repeated in a given model application. Nonetheless, it has been reported to work agreeably for continuous-error functions such as the root mean square error used in back-propagation neural networks (Cawley and Talbot, 2004).

Methods that attempt to approximate bias and variability of an estimator by using values on subsamples are called resampling methods. In jackknife resampling, the bias is estimated by systematically removing one datum each time (jackknifed) from the original data set and recalculating the estimator based on the residual samples (Quenouille, 1956). In a neural network application, each time a modeler trains the network, they assimilate into the test set data that have been jackknifed. This results in a separate neural network being tested on each subset of data and trained with all the remaining data. For jackknife resampling (and cross-validation) to be effective, comprehensive knowledge about the error distribution must be known (Priddy and Keller, 2005). In the absence of this knowledge, a normal distribution is assumed, and bootstrap resampling is used to deal with the small sample size. Efron (1979) explained that in this absence, the sample data set itself offers a representative guide to the sampling distribution. Bootstrap samples are usually generated by replacement sampling from the primary data. For a set of n points, a distinct point has probability $n = 1$ of being chosen on each draw. Modelers can then use the bootstrap samples to construct an empirical distribution of the estimator. Monte Carlo simulation methods can be used to acquire bootstrap resampling results by randomly generating new data sets to simulate the process of data generation. The bootstrap set generates the data used to train a neural network, with the remaining data being used for testing purposes. The bootstrap resampling method is simplistic in application and can be harnessed to derive estimators of standard errors and confidence intervals for multi-faceted estimators of complex parameters of the distribution.

3.6 NEURAL NETWORK TRAINING

A properly trained neural network is one that has "learned" to distinguish patterns derived from input variables and their associated outputs, and affords superior predictive accuracy for an extensive assortment of applications. Neural network connection strengths are adjusted iteratively according to the exhibited prediction error with improved performance driven by sufficient and properly processed input data fed into the model, and a correctly defined learning rule. Two distinct learning paradigms, supervised and unsupervised, will be covered in detail in subsequent sections. A third paradigm, reinforcement learning, is considered to be an intermediate variety of the foregoing two types. In this type of learning, some feedback from the environment is given, but such an indicator is often considered only evaluative, not instructive (Sutton and Barto, 1998). Work by Kaelbling et al. (1996) and Schmidhuber (1996) describe the main strategies for solving reinforcement learning problems. In this approach, the learner collects feedback about the appropriateness of its response. For accurate responses, reinforcement learning bears a resemblance to supervised learning; in both cases, the learner receives information that what it did was, in fact, correct.

3.6.1 LEARNING RULES

Neural network learning and organization are essential for comprehending the neural network architecture discussion in Chapter 2, with the choice of a learning algorithm being central to proper network development. At the center of this development are interconnection weights that allow structural evolution for optimum computation (Sundareshan et al., 1999). The process of determining a set of connection weights to carry out this computation is termed a learning rule. In our discussion, numerous modifications to Hebb's rule, and rules inspired by Hebb, will be covered, given the significant impact of his work on improving the computational power of neural networks. Hebb reasoned that in biological systems, learning proceeds via the adaptation of the strengths of the synaptic interactions between neurons (Hebb, 1949). More specifically, if one neuron takes part in firing another, the strength of the connection between them will be increased. If an input elicits a pattern of neural activity, Hebbian learning will likely strengthen the tendency to extract a similar pattern of activity on subsequent occasions (McClelland et al., 1999). The Hebbian rule computes changes in connection strengths, where pre- and postsynaptic neural activities dictate this process (Rădulescu et al., 2009). A common interpretation of Hebbian rules reads as follows:

$$\Delta W_{ij} = rx_j x_i \qquad (3.6)$$

where x_j = the output of the presynaptic neuron, x_i = the output of the postsynaptic neuron, w_{ij} = the strength of the connection between the presynaptic and postsynaptic neurons, and r = the learning rate. Fundamentally, the learning rate is used to adjust the size of the weight changes. If r is too small, the algorithm will take extended time to converge. Conversely, if r is too large, the algorithm diverges about the error surface (Figure 3.4). Determining the appropriate learning rate is typically achieved through a series of trial and error experiments. Hebbian learning applies to both supervised and unsupervised learning with discussion provided in subsequent sections. A detailed discussion on Hebbian errors in learning, and the Oja modification (Oja, 1982), can be found in a number of dedicated sources provided in the literature (e.g., Rădulescu et al., 2009).

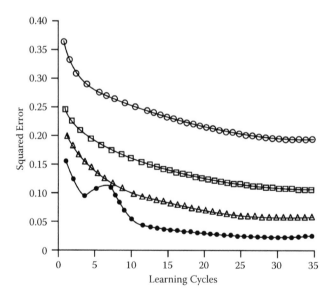

FIGURE 3.4 Short-term view of squared error for an entire training set over several learning cycles [learning rates = 0.05 (circle), 0.10 (square), 0.5 (triangle), and 1.0 (darkened circle)]. Increasing the learning rate hastens the arrival at lower positions on the error surface. If it is too large, the algorithm diverges about the surface.

3.6.2 SUPERVISED LEARNING

Supervised learning has been the mainstream of neural network model development and can be formulated as a nonlinear optimization problem, where the network parameters are the independent variables to be attuned, and an error measure acting as the dependent variable (dos Santos and Von Zuben, 2000). The network is trained (Figure 3.5) by providing it with input and desired outputs ("training set"). Here, the network is adjusted based on careful comparison of the output and the target until the network output matches the specified target. In this process, an output value is produced, and each observation from the training set is processed through the network. The output value is then compared to the actual value (target variable), and the prediction error is computed. Typically, neural network models measure this fit by the use of the mean square error (*MSE*) given by

$$MSE = \frac{SSE}{n} \qquad (3.7)$$

where SSE is the sum of squared error given by

$$SSE = \text{Sum}_{(i=1 \text{ to } n)}\{w_i\,(y_i - f_i)^2\} \qquad (3.8)$$

where y_i = the observed data value, f_i = the predicted value from the fit, and w_i = the weighting applied to each data point, typically, $w_i = 1$. The ultimate goal is to therefore assemble a set of modeled weights that will minimize *MSE* when used as a measurement of fit. However, if sigmoid functions are used, no closed-form solution for minimizing *MSE* exists (Larose, 2004). Nonlinear optimization meth-

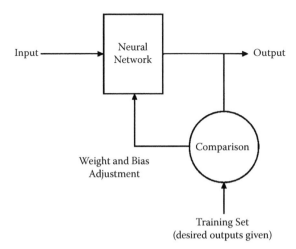

FIGURE 3.5 Supervised learning network with the training data consisting of many pairs of input/output training patterns. Desired outputs given.

ods are therefore needed to facilitate finding a set of weights that will allow the minimization of *MSE*.

3.6.2.1   The Perceptron Learning Rule

Recall our discussion on single-layer perceptrons in Section 2.2.1. The simple perceptron training rule developed by Rosenblatt (1962) is effective if the training examples are linearly separable:

$$w_{ij} \rightarrow w_{ij} + \Delta w_{ij} \tag{3.9}$$

with
$$\Delta w_{ij} = r(x_{Ti} - x_{Oi})x_{j} \tag{3.10}$$

where x_{Ti} = the target output of neuron i, x_{Oi} = the neuron's actual output, x_{j} = the state of the input neuron j, and r ($r > 0$) = the learning rate. If the target output is equivalent to the computed output, $X_{Ti} = X_{Oi}$, then the weight vector of the ith output node remains unaffected. The learning process terminates when all weight vectors w remains unchanged during an inclusive training sequence. The perceptron rule is guaranteed to converge to a solution in an infinite number of steps, so long as a solution exists (Hagan et al., 1999). If, however, perceptron learning is run on a nonseparable set of training examples (see the XOR problem discussed in Chapter 2), the algorithm will not behave properly. In other words, the perceptron learning algorithm, even if terminated after a considerable number of iterations, confers no guarantee as to the quality of the weight produced (Gallant, 1993).

3.6.2.2   Gradient Descent and Back-Propagation

When considering multilayer perceptrons, the error surface is a nonlinear function of weight vector *w*. Algorithms that approximate by gradient descent are employed to train MLPs. To train by the gradient decent algorithm, the gradient *G* of the cost function with respect to each weight w_{ij} of the network must be computable. This process gives information on how each change in weight will affect the overall error *E*. Here, we can consider the concept of *MSE* subsequently described and construct the vector derivative (or gradient):

$$\nabla MSE = \left[\frac{\partial MSE}{\partial w_0}, \frac{\partial MSE}{\partial w_1}, ..., \frac{\partial MSE}{\partial w_m} \right] \tag{3.11}$$

at any position in the weight space. If investigators are given a certain weight vector *w*, and would like to search for another vector *w** with lower *E*, stepping in the direction of the negative gradient is prudent, as this is the direction along which *E* decreases more promptly:

$$w* = w - p\nabla E(w) \tag{3.12}$$

A constant, $p > 0$, regulates how vast a step is completed at each training cycle. As with r, the iteration of Equation 3.6 is termed the gradient descent method of minimizing a function. Note that if the learning rate is selected accurately, the method will likely converge to a local minimum of $E(w)$ for a sufficiently small p, provided that the gradient is nonzero (Gallant, 1993). The size of p can have a substantial effect on the progression of the algorithm, with too small a p resulting in a dramatically protracted approach, and too large a p resulting in a possible oscillation and unsuccessful convergence to a minimum (Gallant, 1993).

Two types of gradient descent learning approaches are encountered: (1) a batch method and (2) an online method. In batch training, the gradient contributions for all training data is accumulated before the weights are updated. In contrast, online training updates weights without delay after the appearance of each training sample. The latter has also been termed stochastic (or noisy), given that single data points can be considered a noisy approximation to the overall gradient G. A "trade-off" approach between batch and online training is termed the mini-batch method, where weight changes are amassed over some number μ of instances before updating the weights (Saad, 1998). Here, $\mu > 1$, but is characteristically smaller than the training set size. There is broad-spectrum consensus among investigators (e.g., Wilson and Martinez, 2003) that the online method is a more resourceful approach for practical gradient descent training situations. Although batch training has been shown to exhibit a considerably less noisy gradient, and has the ability to use optimized matrix operations to compute output and gradient over the whole data set (Saad, 1998), online training has the following advantages (Bishop, 1997; Reed and Marks, 1999):

1. More practical for extensive data sets
2. Faster training, especially when the training set contains many similar data points
3. Gradient noise can aid in escaping from local minima
4. Requires less storage since it does not need to store accumulated weight changes. The only memory of past examples is implicit in the weight vector w.

3.6.2.3 The Delta Learning Rule

Another common learning rule, the delta rule, is regarded as a variant of the Hebbian rule (Rumelhart et al., 1986) and considered a fast and convenient descent algorithm. Developed by Widrow and Hoff (1960), this rule (also termed the least mean square [LMS] rule) attempts to minimize the cumulative error over a given data set as a function of the weights of the network under study:

$$\Delta w_{ij} = rO_j(d_i - O_i) \tag{3.13}$$

where O_j = the output from unit j as input to unit i, d_i = the desired output of unit i, and O_i = the actual output of unit i. This rule implements gradient descent in the sum-squared error for a linear transfer function. It aims to minimize the slope of the cumulative error in a particular region of the network's output function and, hence, it is likely susceptible to local minima. A more involved understanding of the delta rule can be obtained through its derivation (e.g., see Rumelhart et al., 1986, for an explicit example).

3.6.2.4   Back-Propagation Learning Algorithm

The best-known and most widely used supervised learning algorithm is back-propagation. The back-propagation algorithm is used for learning in feedforward networks and was developed to overcome some of the limitations of the perceptron by allowing training for multilayer networks. In order to perform back-propagation, one must minimize E by manipulating weights and employing gradient descent in the weight space to locate the optimal solution (Werbos, 1994). Back-propagation applies the generalized delta learning rule (Klang, 2003; Rumelhart et al., 1986):

$$\Delta w_{ij} = r O_j \left(d_i - O_i \right) f \left(\sum_{j=1}^{n} w_{ij}^{old} O_j \right) \tag{3.14}$$

where $f(\sum_{j=1}^{n} w_{ij}^{old} O_j)$ = the derivative of the sigmoid transfer function (see Chapter 1, Equation 1.4 and Figure 1.5) that monotonically maps its inputs into [0,1]. The full derivatization of this sigmoid function is given by, for example, Gallant (1993). The back-propagation learning algorithm can be decomposed into five defined steps as shown in the following text. More detailed computational descriptions can be found in a variety of literature sources (Gallant, 1993; Chauvin and Rumelhart, 1995; Braspenning et al., 1995). The five steps include the following:

1. **Initialization**—Set all the weights $\{w_{ij}\}$ to small random values and chooses a small positive value for the step size p. This is usually performed on a neuron-by-neuron basis.
2. **Forward-propagation**—Each input is fed to the input layer, and an output O_i is generated based on current weights. O_i is then positioned with d_i by calculating the squared error at each output unit. E is minimized by adjusting the weights appropriately.
3. **Back-propagation**—The total square error computed is propagated back from the output and intermediate layers to the input units.
4. **Weight adjustment**—The weights are updated in the back-propagation. Error gradients in both the output and hidden layers are computed and used in calculating weight corrections and final updates.
5. **Iteration**—Increases the iteration by one, revisits step 2, and repeats the process until the objective function E converges.

The back-propagation method can be exceedingly time consuming in numerous cases, which does not bode well for real-world applications. This slow convergence is due to the plateau phenomenon, which is common in back-propagation-type learning (Saad and Solla, 1995). We shall have more to say on this topic shortly. A number of studies over the past few years have examined ways in which to achieve faster convergence rates while at the same time adhering to the locality constraint. For example, an improper learning rate coefficient can influence the rate of convergence, often resulting in total convergence failure. Jacobs (1988) suggested the use of an adaptive learning rate and that rates should fluctuate over time. The inclusion of a momentum term in Equation 3.6 has also been shown to noticeably increase the rate of convergence. This works by keeping the weight changes on a faster and more directed path by the addition of fractions of previous weight changes (Qian, 1999). Unfortunately, not all of the foregoing alterations have been successful in solving the plateau problem.

Providentially, the gradient (Amari, 1998) and the adaptive natural gradient (Park et al., 2000) methods have been shown to enhance learning speed by avoiding or eliminating plateaus. It has been shown that when a parameter space has a certain underlying structure, the ordinary gradient of a function does not represent its steepest direction (Amari, 1998). Fortunately, the natural gradient does. Both methods are described in detail in Park et al. (2000). In this study, the authors compared both ordinary gradient learning (OGL) and adaptive gradient learning (SANGL) against the well-known bench mark XOR problem previously discussed. Figure 3.6 shows the learning curves for an extended XOR problem using both methods. The pattern set used for training consisted of 1,800 elements of nine clusters, each of which contained 200 elements generated from respective distributions. The test set was similarly generated and consisted of 900 elements. The SANGL method showed more than 250 times faster convergence than OGL in terms of the directed learning cycle. With respect to processing time, the SANGL method was reported to be more than 10 times faster than OGL.

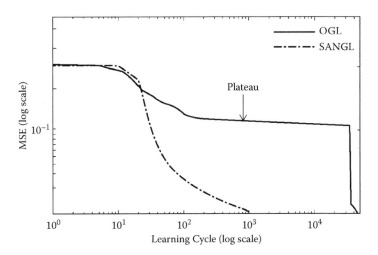

FIGURE 3.6 Learning curves for the extended XOR problem using ordinary gradient learning (OGL) and adaptive natural gradient learning with squared error (SANGL) methods. The plateau problem is clearly visible for the OGL method. The SANGL method achieved a faster convergence speed and thus avoided the plateaus. (Modified from Park et al., 2000. *Neural Networks* 13:755–764. With permission from Elsevier.)

3.6.3   Unsupervised Learning and Self-Organization

In unsupervised learning, the weights and biases are modified in response to network inputs only (Figure 3.7). There are no target outputs available. Ostensibly, the network is trained without benefit of any "teacher" and learns to adapt based on the understanding collected through observed input patterns. The central backbone of all unsupervised learning algorithms is the Hebbian rule. Recall that the Hebbian learning rule states that the connection between two neurons is strengthened if they fire at the same time. Oja subsequently proposed that a similar rule could be used for the unsupervised training of a single nonlinear unit (Oja, 1991). A common feature of unsupervised algorithms is the fact that information is provided by the distance structure between the data points, which is typically determined by the Euclidean metric previously defined in Chapter 2.

3.6.4   The Self-Organizing Map

Among neural network models, the self-organizing map (SOM) is arguably the most commonly employed unsupervised learning algorithm in practice today. Recall from Chapter 2 that the SOM combines two paradigms of unsupervised learning: (1) clustering and (2) projection methods. Practitioners use a self-organizing neural network to create a map in a low-dimensional space, using competitive learning (often termed the winner-take-all approach). Consider a single-layer neural network with inputs $\mathbf{x} \in \Re^d$ fully connected to m outputs. According to the winner-take-all concept, the unit that is closest to the input vector $\mathbf{x}$ takes the prize: $i(\mathbf{x}) = \arg \min_j \|\mathbf{x} - \mathbf{w}_j\|$ (Kohonen, 1984). Learning proceeds with stringent regard to the winner, with the winner having the capacity to adapt its weights fittingly:

$$\Delta \mathbf{w}_{i\,(\mathbf{x})} = \eta(\mathbf{x} - \mathbf{w}_{i\,(\mathbf{x})}) \qquad (3.15)$$

In this process, the neuron whose weight vector was closest to the input vector is updated to be even closer. The winner is therefore more likely to emerge victorious the next time a comparable vector is presented. Upon the presentation of additional

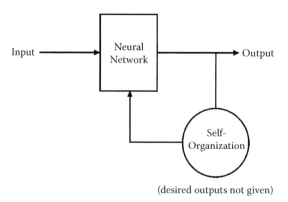

(desired outputs not given)

FIGURE 3.7   Unsupervised learning neural network where the training set consists of input training patterns only. Desired outputs are not given.

inputs, the neuron in the layer closely flanking a group of input vectors soon adjusts its weight vector toward those input vectors. As a result, each cluster of similar input vectors will have a neuron that outputs 1 when a vector in the cluster is displayed, while outputting a 0 at all other times during the process (Sirat and Talbot, 2001).

An attractive aspect of the feature map learning procedure (Equation 3.9) is its ability to be converted to a supervised learning algorithm using Learning Vector Quantization (LVQ). LVQ is a training algorithm for learning classifiers from labeled data samples. In contrast to modeling the class densities, LVQ models the discrimination function defined by the collection of labeled codebook vectors and the nearest-neighborhood search between the codebook and data (de Sa and Ballard, 1993). During the classification process, a given data point x_i is class-assigned based on the class label of the bordering codebook vector. An iterative learning algorithm performing gradient descent updates the winner unit, one defined by (Hollmén et al., 2000):

$$C = \arg\min_k \| x_i - m^k \|$$

(3.16)

Similar to the unsupervised learning algorithms, LVQ relies on the Euclidean metric. As a result, correctly preprocessed and scaled data are essential to ensure that the input dimensions have approximately the same importance for the classification task at hand (Hammer and Vilmann, 2002). For correct data sample classification, where the labels of the winner unit and the data sample are equivalent, the model vector closest to the data sample is attracted toward the sample (Hollmén et al., 2000). Alternatively, if somehow classified incorrectly, the data sample has a repellent effect on the model vector. Full computational details and alternative approaches are covered in more specialized sources (e.g., de Sa and Ballard, 1993; Kohonen, 1995; Hollmén et al., 2000).

3.6.5 BAYESIAN LEARNING CONSIDERATIONS

One of the main technical hitches in neural network model building is dealing with model complexity. Conventional methods of training have just been presented, and as discussed, they can be considered computationally intensive optimization problems based on a set of user-defined training cases. Bayesian learning presents an alternative form of training, an integration problem approach if you will, where the assignment of "prior" probabilities, those based on existing information (e.g., previously collected data, expert judgment), is performed (Lampinen and Vehtari, 2001). Bayes' rules can then be applied to compute the posterior distribution given by

$$P(\theta \mid y) = \frac{f(y|\theta)\pi(\theta)}{\int f(y|\theta)\pi(\theta)d\theta}$$

(3.17)

where θ = the unknown parameter, $\pi(\theta)$ = the prior distribution, y = the experimental data, and $f(y|\theta)$ = the likelihood function. The application of Bayesian methods to neural networks was first reported in the early 1990s (e.g., Buntine and Weigend, 1991;

MacKay, 1992a). Lee (2000) and Lampinen and Vehtari (2001) subsequently detailed their use in MLPs. In the former study, the authors describe the likelihood as

$$
f(y|\theta) = (2\pi\sigma 2)^{n/2} \exp\left[-\frac{1}{2\sigma^2} \sum_{i=1}^{n} \left(y_1 - \beta_0 - \sum_{j=1}^{k} \frac{\beta_j}{1 + \exp\left(-\gamma_{j0} - \sum_{h=1}^{p} \gamma_{jh} x_{ih} \right)} \right)^2 \right]
$$

(3.18)

where n = the sample size, β's and γ's = the weights on the connections, and σ^2 = the variance of the error. This definition is in the context of regression, x, where the inputs are treated as permanent and identified. In such a case, practitioners only need to specify the likelihood for the output, y. A normal distribution for the error termed is assumed. One must also define a network structure, more specifically, the quantity of hidden units in an MLP network, as part of the initial Bayesian learning process. As a framework, Bayesian methodology and Markov chain Monte Carlo (MCMC) sampling are typically employed. In the MCMC process, samples are generated using a Markov chain in the parameter space that exhibits the desired posterior distribution as its stationary distribution (MacKay, 1992b). Moreover, it uses sampled weight vectors from the posterior weight distribution as support of a given neural network structure (Kingston et al., 2005a).

The Bayesian approach has several reported advantages over the frequentist approach. Kingston et al. (2005) refer to it being based on the posterior weight distribution, and hence the lack of need for discovering a single optimum weight vector. They further stress the need for the use of training data unaccompanied as evidence of model evaluation. An independent test set is not needed in most cases. Lee (2000) reports its automatic consideration of all combined model uncertainty into the decisive posterior estimate. Finally, the Bayesian approach provides an effective way to do inference when prior knowledge is wanting or vague (Lampinen and Vehtari, 2001).

3.7 MODEL SELECTION

Model selection is arguably one of the most unsettled issues in neural network design and application. The process of choosing the correct architecture is of primary importance and requires the selection of both the appropriate number of hidden nodes and the connections within (Anders and Korn, 1999). Determining the most appropriate number of hidden nodes depends on a complex array of factors, including (1) the number of input and output units, (2) the number of training patterns, (3) the amount of noise present in the training data, (4) the complexity of the function to be learned, and (5) the training algorithm employed. As previously discussed, choosing too large a number of hidden nodes results in overfitting, leading to poor generalization unless some variety of overfitting prevention (e.g., regularization) is applied. The selection of too few hidden nodes will likely result in high training and generalization errors as a result of underfitting.

There are a variety of reported model selection methods, each with varying degrees of complicatedness and effectiveness. Three general concepts widely reported in the literature include

1. Statistical hypothesis testing
2. Cross-validation and resampling methods
3. Information criteria

Statistical hypothesis testing has been reported as both widely used and reasonably effective in model selection efforts, resulting in networks that closely approximate simulated models (Anders and Korn, 1999). In cross-validation and resampling methods, an estimate of the generalization error can be used as the model selection criterion. In this process, the average of the predictive errors is used to approximate the generalization error. Cross-validation is particularly useful when one has to design an extensive neural network with the goal of good generalization. More established information criteria make explicit use of the number of network weights (or degrees of freedom) in a model (Curry and Morgan, 2006). Some form of measured fit with a penalty term is employed to find the most favorable trade-off between an unbiased approximation of the underlying model and the loss of accuracy caused by parameter estimation (Anders and Korn, 1999).

The Akaike information criterion (AIC) developed by Akaike (1974) can be defined by the following expression:

$$AIC = 2k - 2 \ln(L) \qquad (3.19)$$

where k = the number of free model parameters and L = the maximized value of the likelihood function for the approximated model. Although widely used, the AIC has been reported to be inappropriate for unrealizable models where the number of training samples is small (Murata et al., 1994). Fortunately, one can generalize the AIC with loss criteria, including regularization terms, and study the relation between the training error and the generalization error in terms of the number of the training examples and the complexity of a network. This relationship leads to a network information criterion (NIC), which can then be used to select the optimal network model based on a given training set. The mathematical expression of the NIC can be found in Anders and Korn (1999). The Bayesian approach to model selection termed the Bayesian information criterion (BIC) (Schwarz, 1978) is defined by

$$BIC = -2\log(L) + k\log(n) \qquad (3.20)$$

where n = the number of observations. The BIC is closely related to the AIC but employs Bayesian justification and favors parsimony when assessing model fit. The BIC is reported to allow model averaging for increased predictive performance and better uncertainty accountability (Lee, 2001).

3.8 MODEL VALIDATION AND SENSITIVITY ANALYSIS

Model validation and sensitivity analysis are two concluding steps before models can be used to generate predictions or simulate data. The validation process typically involves subjecting the neural network model to an independent set of validation data that were not utilized for the duration of network training. Various error measures are used to gauge the accuracy of the model, including the MSE described previously in Equations 3.2 and 3.3, with output values compared to target values and the prediction error calculated. Other measures include the root mean square error (RMSE), the mean absolute error (MAE), and the mean relative error (MRE). Kingston et al. (2005b) described the successful use of the coefficient of efficiency, E, in assessing the in-sample and out-of-sample performance of models developed to guarantee physically plausible hydrological modeling. The value of E is calculated by

$$E = 1.0 - \frac{\sum_{i=1}^{n}(O_i - P_i)^2}{\sum_{i=1}^{n}(O_i - \bar{O})^2}$$

(3.21)

where n = the number of data points, O_i = the ith observed data point, $\bar{O}$ = the mean of the observed data, and P_i = the ith predicted data point. The value of E can range from $-\infty$ to 1. Goodness-of-fit measures have also been reported (e.g., May et al., 2009) to describe model performance, including the widely used coefficient of determination, r^2, calculated by

$$r^2 = \frac{\sum_{i=1}^{n}(y_i - \bar{y})(\hat{y}_i - \tilde{y})}{\sqrt{\sum_{i=1}^{n}(y_i - \bar{y})^2 \, n \sum_{i=1}^{n}(\hat{y}_i - \tilde{y})^2}}$$

(3.22)

In general terms, sensitivity analysis is performed to ascertain how susceptible a model is to changes in the value of the parameters and structure of the model. It can be used to assess how the model behavior responds to changes in parameter values during model building and spot relationships determined during training (evaluation). In the latter, the influence of input variables on the dependent variable is examined by evaluating the changes in error committed by the network that would result if an input were removed. As an example application, Pastor-Bárcenas et al. (2005) performed unbiased sensitivity analysis for the development of neural networks for surface ozone modeling. Sensitivity analysis was useful in model development and in identifying the influence of ozone precursors and the ultimate formation of tropospheric ozone.

Investigators have also reported the use of overall connection weights (OCWs) in determining the relative contribution, RC, of each input in predicting the output

(Kingston et al., 2005; May et al., 2009). This is accomplished by dividing each OCW by the sum of the absolute values of all of the OCWs as follows (May et al., 2009):

$$RC_i = \frac{OCW_i}{\sum_{j=1}^{n} |OCW_i|} \times 100 \tag{3.23}$$

where OCW_i is given by

$$OCW_i = \sum_{j=1}^{N_H} f(w_{ij})\upsilon_j \tag{3.24}$$

where f = the transfer function and w_{ij} and υ_j represent the connection weights between layers of the network. As detailed by Kingston et al. (2005), RCs can be compared to prior knowledge of the relationship by which the data were generated in order to assess how well the model has explained the true interactions that take place among the model inputs and outputs.

3.9 CONCLUDING REMARKS

From a practical standpoint, one of the most challenging tasks in developing efficient neural network models is determining the optimum level of complexity to model a chosen problem at hand, particularly given the lack of routine employment of methodical model selection methods. With generalizability being the ultimate aim of model selection, considerations given to the eight-step process outlined in this chapter are paramount, with the modeler's choice of selection criterion being fundamental in providing appropriate predictive abilities for the breadth of applications encountered when studying natural systems. The accurate selection of input variables minimizes the model mismatch error, whereas the selection of a suitable model plays a key role in minimizing model estimation error. Proper training ensures that the neural network has "learned" correctly, recognizing patterns derived from input variables and their associated outputs. Models are then validated by the user to minimize the model prediction error.

REFERENCES

Akaike, H. 1974. A new look at the statistical model identification. *IEEE Transactions on Automatic Control* 19: 716–723.

Amari, S.1998. Natural gradient works efficiently in learning. *Neural Computation* 10: 251–276.

Anders, U., and Korn, O. 1999. Model selection in neural networks. *Neural Networks* 12: 309–323.

Arhonditsis, G.B., 2009. Useless arithmetic? Lessons learnt from aquatic biogeochemical modelling. In G. Hanrahan (Ed.) *Modelling of Pollutants in Complex Environmental Systems*, ILM Publications: St. Albans, UK.

Arhonditsis, G.B., and Brett, M.T. 2004. Evaluation of the current state of mechanistic aquatic biogeochemical modeling. *Marine Ecology: Progress Series* 271: 13–26.

Bailly, K., and Milgram, M. 2009. Boosting feature selection for Neural Network based regression. *Neural Networks* doi:10.1016/j.neunet.2009.06.039.

Bishop, C.M. 1997. *Neural Networks for Pattern Recognition.* Oxford University Press: Oxford.

Braspenning, P.J., Thuijswan, F., and Weijter, A.J.M.M.1995. *Artificial Neural Networks: An Introduction to ANN Theory and Practice.* Springer: Berlin.

Bruce, A., Donoho, D., and Gao, H. 1996. Wavelet analysis. *IEEE Spectrum.* 33: 26–35.

Buntine, W.L., and Weigend, A.S. 1991. Bayesian back-propagation. *Complex Systems* 5: 603–643.

Cai, C., and Harrington, P.B. 1999. Wavelet transform preprocessing for temperature constrained cascade correlation neural networks. *Journal of Chemical Information and Computer Sciences* 39: 874–880.

Cawley, G.C., and Talbot, N.L.C. 2004. Fast leave-one-out cross-validation of sparse least-squares support vector machines. *Neural Networks* 17: 1467–1475.

Chau, F.T., Liang, Y.Z., Gao, J., and Shao, X.G. 2004. *Chemometrics: From Basics to Wavelet Transform.* John Wiley & Sons: Hoboken, NJ.

Chauvin, Y., and Rumelhart, D.E. 1995. *Backpropagation: Theory, Architectures, and Applications.* Lawrence Erlbaum: Philadelphia.

Cooper, G.R.J., and Cowan, D.R. 2008. Comparing time series using wavelet-based semblance analysis. *Computers and Geosciences* 34: 95–102.

Curry, B., and Morgan, P.H. 2006. Model selection in neural networks: Some difficulties. *European Journal of Operational Research* 170: 567–577.

de Sa, V.R., and Ballard, D.H. 1993. A Note on Learning Vector Quantization. In C.L. Giles, S.J. Hanson, and J.D. Cowen (Eds.), *Advances in Neural Information Processing Systems 5*, pp. 220–227. Morgan Kaufmann: San Francisco, CA.

Dorn, E.D., McDonald, G.D., Storrie-Lombardi, M.C., and Nealson, K.H. 2003. Principal component analysis and neural networks for detection of amino acid biosignatures. *Icarus* 166: 403–409.

dos Santos, E.P., and Von Zuben, F.J. 2000. Efficient second-order learning algorithms for discrete-time recurrent neural networks. In L.R. Medsker and L.C. Jain (Eds.), *Recurrent Neural Networks: Design and Applications.* CRC Press, Boca Raton.

Efron, B. 1979. Bootstrap methods: Another look at the jackknife. *The Annals of Statistics* 7: 1–26.

Gallant, S.I. 1993. *Neural Network Learning and Expert Systems.* MIT Press: Cambridge.

Gill, M.K., Asefa, T., Kaheil, Y. And McKee, M. 2007. Effect of missing data on performance of learning algorithms for hydrologic predictions: Implications to an imputation technique. Water Resources Research 43: W07416, doi:10.1029/2006WR005298.

Granger, C.W.J., and Siklos, P.L. 1995. Systematic sampling, temporal aggregation, seasonal adjustment, and cointegration theory and evidence. *Journal of Environmetrics* 66: 357–369.

Hagan, M.T., Demuth, H.B., and Beale, M.H. 1999. *Neural Network Design.* PWS Publishing Company: Boston.

Hammer, B., and Vilmann, T. 2002. Generalized relevance learning vector quantization. *Neural Networks* 15: 1059–1068.

Hanrahan, G. 2010. Computational neural networks driving complex analytical problem solving. *Analytical Chemistry,* 82: 4307–4313.

Hebb, D.O. 1949. *The Organization of Behaviour.* John Wiley & Sons: New York.

Hodge, V.J., Lees, K.J., and Austin, J.L. 2004. A high performance *k*-NN approach using binary neural networks. *Neural Networks* 17: 441–458.

Hollmén, J., Tresp, V., and Simula, O. 2000. A Learning Vector Quantization Algorithm for Probalistic Models. In *Proceedings of EUSIPCO 2000*, 10th European Signal Processing Conference, Volume II, pp. 721–724.

Jacob, C., Kent, A.D., Benson, B.J., Newton, R.J., and McMahon, K.D. 2005. Biological databases for linking large microbial and environmental data sets. In *Proceedings of the 9th World Multiconference on Systemics, Cybernetics and Informatics.* Vol. VI: 279–282. July 10–13, 2005, Orlando, FL.

Jacobs, R.A. 1988. Increased rates of convergence through learning rate adaptation. *Neural Networks* 1: 295–307.

Kaelbling, L.P., Littman, M.L., and Moore, A.W. 1996. Reinforcement learning: A survey. *Journal of Artificial Intelligence Research* 4: 237–285.

Kingston, G.B., Maier, H.R., and Lambert, M.F. 2005a. *A Bayesian Approach to Artificial Neural Network Model Selection.* MODSIM 2005 International Congress on Modelling and Simulation: Modelling and Simulation Society of Australia and New Zealand, December 2005/Andre Zerger and Robert M. Argent (Eds.): pp.1853–1859.

Kingston, G.B., Maier, H.R., and Lambert, M.F. 2005b. Calibration and validation of neural networks to ensure physically plausible hydrological modeling. *Journal of Hydrology* 314: 158–176.

Klang, M. 2003. Neural networks. In *Encyclopedia of Information Systems, Volume 3.* Elsevier Science: Amsterdam. p. 311.

Kohonen, T. 1984. *Self-Organization and Associative Memory.* Springer-Verlag: Berlin.

Kohonen, T. 1995. *Self-Organizing Maps.* Springer-Verlag: New York.

Kumar, P., and Foufoula-Georgiou, E. 1997. Wavelet analysis for geophysical applications. *Reviews of Geophysics* 35: 385–412.

Lampinen, J., and Vehtari, A. 2001. Bayesian approach for neural networks: Review and case studies. *Neural Networks* 14: 257–274.

Larose, D.T. 2004. *Discovering Knowledge in Data: An Introduction to Data Mining.* John Wiley & Sons: Hoboken, NJ.

Lee, H.K.H. 2000. Consistency of posterior distributions for neural networks. *Neural Networks* 13: 629–642.

Lee, H.K.H. 2001. Model selection for neural networks. *Journal of Classification* 18: 227–243.

Leray, P., and Gallinari, P. 1999. Feature selection with neural networks. *Behaviormetrika* 26: 145–166.

MacKay, D. J. C. 1992a. Bayesian interpolation. *Neural Computation* 4: 415–447.

MacKay, D.J.C. 1992b. A practical Bayesian framework for backpropagation networks. *Neural Computation* 4: 448–472.

May, R.J., Maier, H.R., and Dandy, G.C. 2009. Developing Artificial Neural Networks for Water Quality Modelling and Analysis. In G. Hanrahan (Ed.), *Modelling of Pollutants in Complex Environmental Systems, Volume I.* ILM Publications: St. Albans, U.K.

May, R.J., Maier, H.R., Dandy, G.C., and Gayani Fernando, T.M.K. 2008. Non-linear variable selection for artificial neural networks using partial mutual information. *Environmental Modelling and Software* 23: 1312–1326.

McClelland, J.L., Thomas, A., McCandliss, B.D., and Fiez, J.A. 1999. Understanding failures of learning: Hebbian learning, competition for representational space, and some preliminary experimental data. In J.A. Reggia, E. Ruppin, and D. Glanzman (Eds.), *Disorders of Brain, Behavior and Cognition.* Elsevier: Amsterdam.

Murata, N., Yoshizawa S., and Amari S. 1994. Network information criterion-determining the number of hidden units for an artificial neural network model. *IEEE Transactions on Neural Networks* 6: 865–872.

Oja, E. 1982. A simplified neuron model as a principal component analyzer. *Journal of Mathematical Biology* 15: 267–273.

Oja, E. 1991. Data compression, feature extraction, and autoassociation in feedforward neural networks, *Proceeding of the ICANN-91*, Espoo, Finland, June 24–28, 1991, pp. 737–745.

Özesmi, S.L., Tan, C.O., and U. Özesmi. 2006b. Methodological issues in building, training, and testing artificial neural networks in ecological applications. *Ecological Modelling* 195: 83–93.

Özesmi, U., Tan, C.O., Özesmi, S.L., and Robertson, R.J. 2006a. Generalizability of artificial neural network models in ecological applications: Predicting nest occurrence and breeding success of the red-winged blackbird *Agelaius phoeniceus*. *Ecological Modelling* 195: 94–104.

Pantanowitz, A., and Marwala, T. 2009. Missing data imputation through the use of the random forest algorithm. In W. Yu and E.N. Sanchez (Eds.), *Advances in Computational Intelligence*, Springer-Verlag: Berlin, pp. 53–62.

Park, H. Amari, S., and Fukuizu, K. 2000. Adaptive natural gradient learning algorithms for various stochastic models. *Neural Networks* 13: 755–764.

Pastor-Bárcenas, O., Soria-Olivas, E., Martín-Guerrero, J.D., Camps-Valls, G., Carrasco-Rodríguez, J.L., and del Valle-Tascón. 2005. Unbiased sensitivity analysis and pruning techniques in neural networks for surface ozone modelling. *Ecological Modelling* 182: 149–158.

Pelckmans, K., De Brabanter, J., Suykens, J.A.K., and De Moor, B. 2005. Handling missing values in support vector machine classifiers. *Neural Networks* 18: 684–692.

Priddy, K.L., and Keller, P.E. 2005. *Artificial Neural Networks: An Introduction*. SPIE Publications: Bellingham, WA.

Qian, N. 1999. On the momentum term in gradient descent learning algorithms. *Neural Networks* 12: 145–151.

Quenouille, M. H. 1956. Notes on bias in estimation. *Biometrika* 43: 353–360.

Rădulescu, A., Cox, K., and Adams, P. 2009. Hebbian errors in learning: An analysis using the Oja model. *Journal of Theoretical Biology* 258: 489–501.

Reed, R.D., and Marks, R.J. 1999. Neural *Smithing: Supervised Learning in Feedfoward Artificial Neural Networks*. Spartan Press: Washington.

Riveros, T.A., Hanrahan, G., Muliadi, S., Arceo, J., and Gomez, F.A. 2009. On-capillary derivatization using a hybrid artificial neural network-genetic algorithm approach. *Analyst* 134: 2067–2070.

Rosenblatt, F. 1962. *Principles of Neurodynamics: Perceptrons and the Theory of Brain Mechanisms*. Washington: Spartan Books.

Rumelhart, D.E., Hinton, G.E., and McClelland, J.L. 1986. Learning internal representations by error propagation. In D.E. Rumelhart and J.L. McClelland (Eds.), *Parallel Distributed Processing: Explorations in the Microstructure of Cognition. Volume I*. MIT Press: Cambridge, MA.

Saad, D., 1998. *On-line Learning in Neural Networks*. Cambridge University Press: Cambridge.

Saad, D., and Solla, S.A. 1995. On-line learning in soft committee machines. *Physical Review E* 52: 4225–4243.

Samani, N., Gohari-Moghadam, M., and Safavi, A.A. 2007. A simple neural network model for the determination of aquifer parameters. *Journal of Hydrology* 340: 1–11.

Sanders, J.A., Morrow-Howell, N., Spitznagel, E., Doré, P., Proctor, E.K., and Pescarino, R. 2006. Imputing missing data: A comparison of methods for social work researchers. *Social Work Research* 30: 19–31.

Schmidhuber, J. 1996. A general method for multi-agent learning and incremental self-improvement in unrestricted environments. In X. Yao, (Ed.), *Evolutionary Computation: Theory and Applications*. Scientific Publishing Company: Singapore, pp. 84–87.

Schwarz, G. 1978. Estimating the dimension of a model. *The Annals of Statistics* 6: 461–464.

Shrestha, D.L., Kayastha, N., and Solomatine, D.P. 2009. A novel approach to parameter uncertainty analysis of hydrological models using neural networks. *Hydrology and Earth System Sciences* 13: 1235–1248.

Sirat, M., and Talbot, C.J. 2001. Application of artificial neural networks to fracture analysis at the Äspo HRL, Sweden: A fracture set. *International Journal of Rock Mechanics and Mining Sciences* 38: 621–639.

Sundareshan, M.K., Wong, Y.C., and Condarcure, T. 1999. Training algorithms for recurrent neural nets that eliminate the need for computation of error gradients with applications to trajectory production problem. In L.R. Medsker and L.C. Jain (Eds.), *Recurrent Neural Networks: Design and Applications*. CRC Press: Boca Raton, FL.

Sutton, R.S., and Barto, A.G. 1998. *Reinforcement Learning: An Introduction*. MIT Press: Cambridge, MA.

Tranter, R. 2000. *Design and Analysis in Chemical Research*. CRC Press: Boca Raton, FL.

Uncu, O., and Türkşen, I.B. 2007. A novel feature selection approach: Combining feature wrappers and filters. *Information Sciences* 177: 449–466.

Werbos, P.J.1994. *The Roots of Backpropagation: From Ordered Derivatives to Neural Networks and Political Forecasting*. John Wiley & Sons: Hoboken.

Widrow, B., and Hoff, M.E. 1960. Adaptive switching circuits. *IRE WESCON Convention Record* 4: 96–104.

Wilson, D.R., and Martinez, T.R. 2003. The general inefficiency of batch training for gradient descent learning. *Neural Networks* 16: 1429–145

Wolpert, D.H. 1996. The lack of a priori distinctions between learning algorithms, *Neural Computation* 8: 1341–1390.

4 Intelligent Neural Network Systems and Evolutionary Learning

4.1 HYBRID NEURAL SYSTEMS

Although several conventional methods (e.g., pruning techniques) exist that may be used to automatically determine neural network configuration and weights, they are often susceptible to trapping at local optima and characteristically dependent on the initial network structure (Gao et al., 1999). Overcoming these limitations would thus prove useful in the development of more intelligent neural network systems, those capable of demonstrating effective global search characteristics with fast convergence, and capable of providing alternative tools for modeling complex natural processes. The movement toward more intelligent network systems requires consideration of alternative strategies to help optimize neural network structure and aid in bringing multifaceted problems into focus. By converging on specific methodologies common to, for example, genetic algorithms, evolutionary programming, and fuzzy logic, we can begin to comprehend how when combined with neural networks, hybrid technology can impart the efficiency and accuracy needed in fundamental research, where multidisciplinary and multiobjective tasks are routinely performed. The objectives of this chapter are twofold: (1) to present theoretical concepts behind unconventional methodologies and (2) showcase a variety of example hybrid techniques, including neuro-fuzzy, neuro-genetic, and neuro-fuzzy-genetic systems.

4.2 AN INTRODUCTION TO GENETIC ALGORITHMS

One of the most popular hybrid evolutionary models involves the incorporation of genetic algorithms (GAs). A genetic algorithm is a highly parallel, randomly searching algorithm that is said to emulate evolution according to the Darwinian survival of the fittest principle, with better solutions having a higher probability of being selected for reproduction than poorer solutions (Goldberg, 1989). I am reminded of Darwin's *On the Origin of Species*, in which a noteworthy passage reads:

> It may be said that natural selection is daily and hourly scrutinizing, throughout the world, every variation, even the slightest; rejecting that which is bad, preserving and adding up all that is good; silently and insensibly working, whenever and wherever opportunity offers, at the improvement of each organic being in relation to its organic and inorganic conditions of life. We see nothing of these slow changes in progress,

until the hand of time has marked the long lapses of ages, and then so imperfect is our view into long past geological ages, that we only see that the forms of life are now different from what they formerly were.

In his own work, Holland elucidated the adaptive process of natural systems and outlined the two main principles of GAs: (1) their ability to encode complex structures through bit-string representation and (2) complex structural improvement via simple transformation (Holland, 1975). Unlike the gradient descent techniques discussed in Chapter 3, the genetic algorithm search is not biased toward locally optimal solutions (Choy and Sanctuary, 1998). The basic outline of a traditional GA is shown in Figure 4.1, with the rather simple mechanics of this basic approach highlighted in the following text. As depicted, a GA is an iterative procedure operating on a population of a given size and executed in a defined manner. Although there are many possible variants on the basic GA, the operation of a standard algorithm is described by the following steps:

1. **Population initialization:** The random formation of an initial population of chromosomes with appropriate encoding of the examples in the problem domain to a chromosome.
2. **Fitness evaluation:** The fitness $f(x)$ of each chromosome x in the population is appraised. If the optimal solution is obtained, the algorithm is stopped

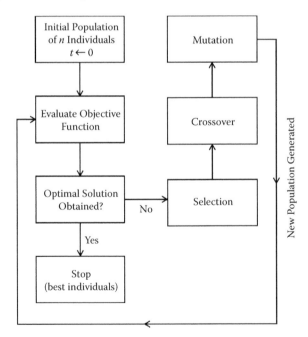

FIGURE 4.1   A generalized genetic algorithm outline. The algorithm consists of population initialization, fitness evaluation, selection, crossover, mutation, and new population evaluation. The population is expected to converge to optimal solutions over iterations of random variation and selection.

with the best individuals chosen. If not, the algorithm proceeds to the selection phase of the iterative process.

3. **Selection:** Two parent chromosomes from a population are selected according to their fitness. Strings with a higher fitness value possess an advanced probability of contributing to one or more offspring in the subsequent generation.

4. **Crossover:** Newly reproduced strings in the pool are mated at random to form new offspring. In single point crossover, one chooses a locus at which to swap the remaining alleles from one parent to the other. This is shown visually in subsequent paragraphs.

5. **Mutation:** Alteration of particular attributes of new offspring at a locus point (position in an individual chromosome) with a certain probability. If no mutation occurs, the offspring is the direct result of crossover, or a direct copy of one of the parents.

6. **New Population Evaluation:** The use of a newly generated population for an additional run of the algorithm. If the end condition is satisfied, stopped, and returned, the best solution in the current population is achieved.

The way in which operators are used—and the representation of the genotypes involved—will dictate how a population is modeled. The evolving entities within a GA are repeatedly referred to as genomes, whereas related Evolving Strategies (ES) model the evolutionary principles at the level of individuals or phenotypes (Schwefel and Bäck, 1997). Their most important feature is the encoding of so-called strategic parameters contained by the set of individual characters. They have achieved widespread acceptance as robust optimization algorithms in the last two decades and continue to be updated to suit modern-day research endeavors. This section will concentrate solely on GAs and their use in understanding adaptation phenomena in modeling complex systems. More detailed coverage of the steps in a GA process is given in the following text.

4.2.1 INITIATION AND ENCODING

A population of n chromosomes (possible solutions to the given problem) is first created for problem solving by generating solution vectors within the problem space: a space for all possible reasonable solutions. A position or set of positions in a chromosome is termed a gene, with the possible values of a gene known as alleles. More specifically, in biological systems, an allele is an alternative form of a gene (an individual member of a pair) that is situated at a specific position on an identifiable chromosome. The fitness of alleles is of prime importance; a highly fit population is one that has a high reproductive output or has a low probability of becoming extinct. Similarly, in a GA, each individual chromosome has a fitness function that measures how fit it is for the problem at hand. One of the most critical considerations in applying a GA is finding a suitable encoding of the examples in the problem domain to a chromosome, with the type of encoding having dramatic impacts on evolvability, convergence, and overall success of the algorithm (Rothlauf, 2006). There are four commonly employed encoding methods used in GAs: (1) binary encoding, (2) permutation encoding, (3) value encoding, and (4) tree encoding.

4.2.1.1   Binary Encoding

Binary encoding is the most common and simplest form of encoding used. In this process, every chromosome is a string of bits, 0 or 1 (e.g., Table 4.1). As a result, a chromosome is a vector x consisting of l genes c_i:

$$x = (c_1, c_2, ...c_l) \quad c_l \in \{0,1\}$$

where l = the length of the chromosome. Binary encoding has been shown to provide many possible chromosomes even with a small number of alleles. Nevertheless, much of the traditional GA theory is based on the assumption of fixed-length, fixed-order binary encoding, which has proved challenging for many problems, for example, evolving weights for neural networks (Mitchell, 1998). Various modifications (with examples provided in the following sections) have been recently developed so that it can continue to be used in routine applications.

4.2.1.1.1   Permutation Encoding

In permutation encoding, every chromosome is a string of numbers represented by a particular sequence, for example, Table 4.2. Unfortunately, this approach is limited and only considered ideal for limited ordering problems. Permutation encoding is highly redundant; multiple individuals will likely encode the same solution. If we consider the sequences in Table 4.2, as a solution is decoded from left to right, assignment of objects to groups depends on the objects that have emerged earlier in the chromosome. Therefore, changing the objects encoded at an earlier time in the chromosome may dislocate groups of objects encoded soon after. If a permutation is applied, crossovers and mutations must be designed to leave the chromosome consistent, that is, with sequence format (Sivanandam and Deepa, 2008).

TABLE 4.1

Example Binary Encoding with Chromosomes Represented by a String of Bits (0 or 1)

Chromosome	Bit string
A	10110010110010
B	11111010010111

TABLE 4.2

Example Permutation Encoding with Chromosomes Represented by a String of Numbers (Sequence)

Chromosome	Sequence
A	1 3 4 2 6 5 7 8 9
B	2 4 3 1 7 6 8 5 9

TABLE 4.3

Example Value Encoding with Chromosomes Represented by a String of Real Numbers

Chromosome	Values
A	2.34 1.99 3.03 1.67 1.09
B	1.11 2.08 1.95 3.01 2.99

4.2.1.1.2 Direct Value Encoding

In direct value encoding, every chromosome is a string of particular values (e.g., form number, real number) so that each solution is encoded as a vector of real-valued coefficients, for example, Table 4.3 (Goldberg, 1991). This has obvious advantages and can be used in place of binary encoding for intricate problems (recall our previous discussion on evolving weights in neural networks). More specifically, in optimization problems dealing with parameters with variables in continuous domains, it is reportedly more intuitive to represent the genes directly as real numbers since the representations of the solutions are very close to the natural formulation; that is, there are no differences between the genotype (coding) and the phenotype (search space) (Blanco et al., 2001). However, it has been reported that use of this type of coding often necessitates the development of new crossover and mutation specific for the problem under study (Hrstka and Kučerová, 2004).

4.2.1.1.3 Tree Encoding

Tree encoding is typically used for genetic programming, where every chromosome is a tree of some objects, for example, functions or commands in a programming language (Koza, 1992). As detailed by Schmidt and Lipson (2007), tree encodings characteristically define a root node that represents the final output (or prediction) of a candidate solution. Moreover, each node can have one or more offspring nodes that are drawn on to evaluate its value or performance. Tree encodings (e.g., Figure 4.2) in symbolic regression are termed expression trees, with evaluation invoked by calling the root node, which in turn evaluates its offspring nodes. Recursion stops at the terminal nodes, and evaluation collapses back to the root (Schmidt and Lipson, 2007). The problem lies in the potential for uncontrolled growth, preventing the formation of a more structured, hierarchical candidate solution (Koza, 1992). Further, the resulting trees, if large in structure, can be difficult to understand and simplify (Mitchell, 1998).

4.2.2 Fitness and Objective Function Evaluation

The fitness of an individual GA is the value of an objective function for its phenotype (Sivanandam and Deepa, 2008). Here, the fitness $f(x)$ of each chromosome x in the population is evaluated. More specifically, the fitness function takes one individual from a GA population as input and evaluates the encoded solution of that particular individual. In essence, it is a particular type of objective function that quantifies the optimality of a solution by returning a fitness value that denotes how good a solution

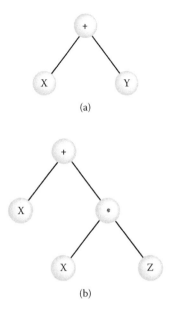

FIGURE 4.2 Tree encodings. Example expressions: (a) $f(x) = x + y$ and (b) $x + (x*z)$.

this individual is. In general, higher fitness values represent enhanced solutions. If the optimal solution is obtained, the algorithm is stopped with the best individuals chosen. If not, the algorithm proceeds to the selection phase.

Given the inherent difficulty faced in routine optimization problems, for example, constraints on their solutions, fitness functions are often difficult to ascertain. This is predominately the case when considering multiobjective optimization problems, where investigators must determine if one solution is more appropriate than another. Individuals must also be aware that in such situations not all solutions are feasible. Traditional genetic algorithms are well suited to handle this class of problems and accommodate multiobjective problems by using specialized fitness functions and introducing methods to promote solution diversity (Konak et al., 2006). The same authors detailed two general approaches to multiobjective optimization: (1) combining individual objective functions into a single composite function or moving all but one objective to the constraint set, and (2) determination of an entire Pareto optimal solution set (a set of solutions that are nondominated with respect to each other) or a representative subset (Konak et al., 2006). In reference to Pareto solutions, they are qualified in providing the decision maker with invaluable insight into the multiobjective problem and consequently sway the derivation of a best decision that can fulfill the performance criteria set (Chang et al., 1999).

4.2.3 Selection

The selection of the next generation of chromosomes is a random process that assigns higher probabilities of being selected to those chromosomes with superior fitness values. In essence, the selection operator symbolizes the process of natural selection in biological systems (Goldberg, 1989). Once each individual has been evaluated, individuals with the highest fitness functions will be combined to produce a second generation. In general, the second generation of individuals can be expected to be "fitter" than the first, as it was derived only from individuals carrying high fitness functions. Therefore, solutions with higher objective function values are more likely to be convincingly chosen for reproduction in the subsequent generation.

Two main types of selection methods are typically encountered: (1) fitness proportionate selection and (2) rank selection. In fitness proportionate selection, the probability of a chromosome being selected for reproduction is proportionate to its fitness value (Goldberg, 1989). The most common fitness proportionate selection technique is termed roulette wheel selection. Conceptually, each member of the population is allocated a section of an imaginary roulette wheel, with wheel sections proportional to the individual's fitness (e.g., the fitter the individual, the larger the section of the wheel it occupies). If the wheel were to be spun, the individual associated with the winning section will be selected. In rank selection, individuals are sorted by fitness and the probability that an individual will be selected is proportional to its rank in the sorter list. Rank selection has a tendency to avoid premature convergence by alleviating selection demands for large fitness differentials that occur in previous generations (Mitchell, 1998).

4.2.4   CROSSOVER

Once chromosomes with high fitness values are selected, they can be recombined into new chromosomes in a procedure appropriately termed *crossover*. The crossover operator has been consistently reported to be one of the foremost search operators in GAs due to the exploitation of the available information in previous samples to influence subsequent searches (Kita, 2001). Ostensibly, the crossover process describes the process by which individuals breed to produce offspring and involves selecting an arbitrary position in the string and substituting the segments of this position with another string partitioned correspondingly to produce two new offspring (Kellegöz et al., 2008). The crossover probability (P_c) is the fundamental parameter involved in the crossover process. For example, if $P_c = 100\%$, then all offspring are constructed by the crossover process (Sivanandam and Deepa, 2008). Alternatively, if $P_c = 0\%$, then a completely different generation is constructed from precise copies of chromosomes from an earlier population.

In a single-point crossover (Figure 4.3), one crossover point is selected, and the binary string from the beginning of the chromosome to the crossover point is copied from one parent with the remaining string copied from the other parent. The two-point crossover operator differs from the one-point crossover in that the two crossover points are selected randomly. More specifically, two crossover points are selected with the binary string from the beginning of chromosome to the first crossover point copied from one parent. The segment from the first to the second crossover point is copied from the second parent, and the

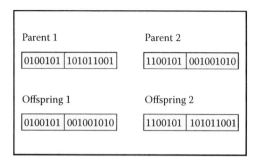

Parent 1   Parent 2

| 0100101 | 101011001 | | 1100101 | 001001010 |

Offspring 1   Offspring 2

| 0100101 | 001001010 | | 1100101 | 101011001 |

FIGURE 4.3 *A color version of this figure follows page 106.* Illustration of the single-point crossover process. As depicted, the two parent chromosomes are cut once at corresponding points and the sections after the cuts swapped with a crossover point selected randomly along the length of the mated strings. Two offspring are then produced.

rest is copied from the first parent. Multipoint crossover techniques have also been employed, including n-point crossover and uniform crossover. In n-point crossover, n cut points are randomly chosen within the strings and the $n - 1$ segments between the n cut points of the two parents are exchanged (Yang, 2002). Uniform crossover is a so-called generalization of n-point crossover that utilizes a random binary vector, which is the same length of the parent chromosome, which creates offspring by swapping each bit of two parents with a specified probability (Syswerda, 1989). It is used to select which genes from each parent should be crossed over. Note that if no crossover is performed, offspring are precise reproductions of the parents.

4.2.5 MUTATION

The mutation operator, while considered secondary to selection and crossover operators, is a fundamental component to the GA process, given its ability to overcome lost genetic information during the selection and crossover processes (Reid, 1996). As expected, there are a variety of different forms of mutation for the different kinds of representation. In binary terms, mutations randomly alter (according to some probability) some of the bits in the population from 1 to 0 or vice versa. The objective function outputs allied with the new population are calculated and the process repeated. Typically, in genetic algorithms, this probability of mutation is on the order of one in several thousand (Reid, 1996). Reid also likens the mutation operator to an adaptation and degeneration of crossover; an individual is crossed with a random vector, with a crossover segment that consists only of the chosen allele. For this reason, he claims that the justification of the search for a feasible mutation takes a similar form to that of feasible crossover. Similar to the crossover process, mutation is assessed by a probability parameter (P_m). For example, if $P_m = 100\%$, then the whole chromosome is ostensibly altered (Sivanandam and Deepa, 2008). Alternatively, if $P_m = 0\%$, then nothing changes.

4.3 AN INTRODUCTION TO FUZZY CONCEPTS AND FUZZY INFERENCE SYSTEMS

The basic concepts of classical set theory in mathematics are well established in scientific thought, with knowledge expressed in quantitative terms and elements either belonging exclusively to a set or not belonging to a set at all. More specifically, set theory deals with sets that are "crisp," in the sense that elements are either in or out according to rules of common binary logic. Ordinary set-theoretic representations will thus require the preservation of a crisp differentiation in dogmatic fashion. If we reason in terms of model formation, uncertainty can be categorized as either "reducible" or "irreducible." Natural uncertainty is irreducible (inherent), whereas data and model uncertainty include both reducible and irreducible constituents (Kooistra et al., 2005). If the classical approach is obeyed, uncertainty is conveyed by sets of jointly exclusive alternatives in circumstances where one alternative is favored. Under these conditions, uncertainties are labeled as diagnostic, predictive, and retrodictive, all arising from nonspecificity inherent in each given set (Klir and Smith, 2001). Expansion of the formalized language of classical set theory has led to two important generalizations in the field of mathematics: (1) fuzzy set theory and (2) the theory of monotone measures. For our discussion, concentration on fuzzy set theory is of prime interest, with foundational concepts introduced and subsequently expanded upon by Lotfi Zadeh (Zadeh, 1965, 1978). A more detailed historical view of the development of mathematical fuzzy logic and formalized set theory can be found in a paper by Gottwald (2005). This development has substantially enlarged the framework for formalizing uncertainty and has imparted a major new paradigm in the areas of modeling and reasoning, especially in the natural and physical sciences.

4.3.1 FUZZY SETS

Broadly defined by Zadeh (1965), a fuzzy set is a class of objects with a continuum of grades of "membership" that assigns every object a condition of membership ranging between zero and one. Fuzzy sets are analogous to the classical set theory framework, but do offer a broader scale of applicability. In Zadeh's words:

> Essentially, such a framework provides a natural way of dealing with problems in which the source of imprecision is the absence of sharply defined criteria of class membership rather than the presence of random variables.

Each membership function, denoted by

$$\mu_A (x): X \rightarrow [0, 1]$$

defines a fuzzy set on a prearranged universal set by assigning to each element of the universal set its membership grade in the fuzzy set. A is a standard fuzzy set, and X is the universal set under study. The principal term μ_A is the element X's degree of membership in A. The fuzzy set allows a continuum of probable choices, for example:

$$\mu_A (x) = 0 \text{ if } x \text{ is } not \text{ in } A \text{ (nonmembership)}$$

$$\mu_A (x) = 1 \text{ if } x \text{ is } entirely \text{ in } A \text{ (complete membership)}$$

$$0 < \mu_A (x) < 1 \text{ if } x \text{ is } partially \text{ in } A \text{ (intermediate membership)}$$

Although a fuzzy set has some resemblance to a probability function when X is a countable set, there are fundamental differences among the two, including the fact that fuzzy sets are exclusively nonstatistical in their characteristics (Zadeh, 1965). For example, the grade of membership in a fuzzy set has nothing in common with the statistical term *probability*. If probability was to be considered, one would have to study an exclusive phenomenon, for example, whether it would or would not actually take place. Referring back to fuzzy sets, however, it is possible to describe the "fuzzy" or indefinable notions in themselves. As will be evident, the unquestionable preponderance of phenomena in natural systems is revealed simply by imprecise perceptions that are characterized by means of some rudimentary form of natural language. As a final point, and as will be revealed in the subsequent section of this chapter, the foremost objective of fuzzy sets is to model the semantics of a natural language; hence, numerous specializations in the biological and environmental sciences will likely exist in which fuzzy sets can be of practical importance.

4.3.2 Fuzzy Inference and Function Approximation

Fuzzy logic, based on the theory of fuzzy sets, allows for the mapping of an input space through membership functions. It also relies on fuzzy logic operators and parallel IF-THEN rules to form the overall process identified as a fuzzy inference system (Figure 4.4). Ostensibly, fuzzy rules are logical sentences upon which a derivation can be executed; the act of executing this derivation is referred to as an inference process. In fuzzy logic control, an observation of particular aspects of a studied system are taken as input to the fuzzy logic controller, which uses an inference process to delineate a function from the given inputs to the outputs of the controller, thereby changing some aspects of the system (Brubaker, 1992). Two types of fuzzy inference systems are typically reported in the literature: Mamdani-type inference and Sugeno-type inference models. Mamdani's original investigation (Mamdani, 1976) was based on the work of Zadeh (1965), and although his work has been adapted over the years, the basic premise behind this approach has remained nearly unchanged. Mamdani reasoned his fuzzy systems as generalized stochastic systems capable of approximating prescribed random processes with arbitrary accuracy. Although Mamdani systems are more commonly used, Sugeno systems (Sugeno, 1985) are reported to be more compact and computationally efficient (Adly and Abd-El-Hafiz, 2008). Moreover, Sugeno systems are appropriate for constructing fuzzy models

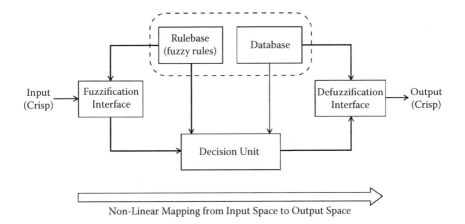

FIGURE 4.4   Diagram of the fuzzy inference process showing the flow input, variable fuzzification, all the way through defuzzification of the cumulative output. The rule base selects the set of fuzzy rules, while the database defines the membership functions used in the fuzzy rules.

based on adaptive techniques and are ideally suited for modeling nonlinear systems by interpolating between multiple linear models (Dubois and Prade, 1999).

In general terms, a fuzzy system can be defined as a set of IF-THEN fuzzy rules that maps inputs to outputs (Kosko, 1994). IF-THEN rules are employed to express a system response in terms of linguistic variables (as summarized by Zadeh, 1965) rather than involved mathematical expressions. Each of the truth-values can be assigned a degree of membership from 0 to 1. Here, the degree of membership becomes important, and as mentioned earlier, is no longer a matter of "true" or "false." One can describe a method for learning of membership functions of the antecedent and consequent parts of the fuzzy IF-THEN rule base given by

$$R_i: \text{IF } x_1 \text{ is } A_{ij} \text{ and } \dots \text{ and } x_n \text{ is } A_{in} \text{ then } y = z_i, \tag{4.1}$$

$i = 1,\dots,m$, where A_{ij} are fuzzy numbers of triangular form and z_i are real numbers defined on the range of the output variable. The membership function is a graphical representation of the magnitude of participation of each input (Figure 4.5). Membership function shape affects how well a fuzzy system of IF-THEN rules approximate a function. A comprehensive study by Zhao and Bose (2002) evaluated membership function shape in detail. Piecewise linear functions constitute the simplest type of membership functions and may be generally of either triangular or trapezoidal type, where the trapezoidal function can take on the shape of a truncated triangle. Let us look at the triangular membership function in more detail. In Figure 4.5a, the a, b, and c represent the x coordinates of the three vertices of $\mu_A(x)$ in a fuzzy set A (a: lower boundary and c: upper boundary, where the membership degree is zero; b: the center, where the membership degree is 1) (Mitaim and Kosko, 2001). Gaussian bell-curve sets have been shown to give richer fuzzy systems with simple learning laws that tune the bell-curve means and variances, but have been

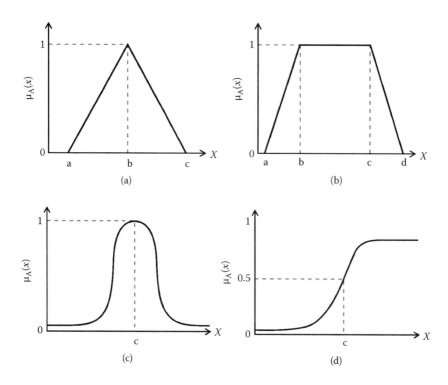

FIGURE 4.5 Example membership functions for a prototypical fuzzy inference system: (a) triangular, (b) trapezoidal, (c) Gaussian, and (d) sigmoid-right. Note the regular interval distribution with triangular functions and trapezoidal functions (gray lines, extremes) assumed. (Based on an original schematic by Adroer et al., 1999. *Industrial and Engineering Chemistry Research* 38: 2709–2719.)

reported to convert fuzzy systems to radial-basis function neural networks or to other well-known systems that predate fuzzy systems (Zhao and Bose, 2002). Yet the debate of which membership function to exercise in fuzzy function approximation continues, as convincingly expressed by Mitiam and Kosko (2001):

> The search for the best shape of if-part (and then-part) sets will continue. There are as many continuous if-part fuzzy subsets of the real line as there are real numbers. ... Fuzzy theorists will never exhaust this search space.

Associated rules make use of input membership values to ascertain their influence on the fuzzy output sets. The fuzzification process can encode mutually the notions of uncertainty and grade of membership. For example, input uncertainty is encoded by having high membership of other likely inputs. As soon as the functions are inferred, scaled, and coalesced, they are defuzzified into a crisp output that powers the system under study. Essentially, defuzzification is a mapping process from a space of fuzzy control actions delineated over an output universe of communication into a space of crisp control actions. Three defuzzification methods are routinely employed: centroid, mean of maxima (MOM), and last of maxima (LOM), where

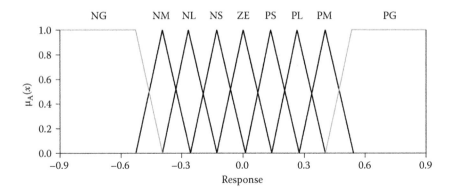

FIGURE 4.6   Illustration of how an output (response) membership function is distributed at regular intervals by the inference process. The overlap between neighboring membership functions provides fine tuning and ensures a response lacking abrupt changes.

the conversion of linguistic values into crisp values is performed (Nurcahyo et al., 2003). The nonlinear mapping from a given input to an output using fuzzy logic then provides a basis from which assessments can be made and/or patterns discerned. Figure 4.6 illustrates output (response) membership functions that have been distributed at regular intervals by the inference process. In essence, there is a 50% overlap between neighboring membership functions (sides of triangles cross each other at a truth value ordinate of 0.5), which provides fine tuning and ensures a response with no abrupt changes (Adroer et al., 1999). Nine example fuzzy sets are visible with linguistic variables defined as NG = negative great, NM = negative medium, NL = negative low, NS = negative small, ZE = zero, PS = positive small, PL = positive low, PM = positive medium, and PG = positive great.

4.3.3   FUZZY INDICES AND EVALUATION OF ENVIRONMENTAL CONDITIONS

Mathematical aspects of fuzzy sets and fuzzy logic have been routinely applied for classifying environmental conditions, and for describing both natural and anthropogenic changes (Silvert, 2000) inexorably linked to natural systems. In his review, Silvert highlighted the ability of fuzzy logic to integrate different kinds of observations in a way that permits a useful balance between favorable and unfavorable observations, and between incommensurable effects such as social, economic, and biological impacts. For example, when developing environmental indices, fuzzy logic displays the ability to combine pertinent indices with much more flexibility than when combining discrete measures, which are ostensibly binary indices corresponding to crisp sets.

A more recent review by Astel (2007) further elaborated on the use of fuzzy logic in investigating the principles of interacting elements/variables and their integration into complex environmental systems. Contemporary environmental research focuses on comprehending the systems, processes, and dynamics that shape the physical, chemical, and biological environment from the molecular to a grander (planetary) scale. Furthermore, human activities and the natural environment are highly intertwined, and human health and well-being are inextricably associated with local, regional, and global ecosystem functioning. Fundamental environmental research integrates and builds on the physical, chemical, biological, and social sciences, mathematics, and engineering. Including fuzzy evaluation in this interdisciplinary approach should therefore be of continued interest and application by practitioners. Astel pinpointed four key advantages of applying fuzzy logic in modern environmental studies. These include

1. The ability to synthesize quantitative information into qualitative output, which is more readable for decision makers and regulators
2. The explicit ability to consider and propagate uncertainties
3. The ability to take on a scalable modular form, which enables easy accommodation to new hazard and exposure parameters
4. The ability to become an effective risk analysis tool for air, water, and soil quality monitoring, given the ease of programmability

As an example, Astel drew attention to a study by Uricchio et al. (2004) that examined a fuzzy knowledge–based decision support system for groundwater pollution risk evaluation. The authors used the combined value of both intrinsic vulnerability of a southern Italian aquifer, obtained by implementing a parametric managerial model (SINTACS), and a degree of hazard value, which considers specific human activities. The designed decision support system took into account the uncertainty of the environmental domain by using fuzzy logic and evaluated the reliability of the results according to information availability. Fuzzy linguistic variables were used to define the degree of hazard for the different factors chosen. For example, by using this approach, they showed that it was possible to diminish the risk that a pollutant may leach along the soil profile and reach groundwater in areas with an intrinsic vulnerability by ranking different anthropic activities according to their degree of hazard. In a related study, Icaga (2007) developed an index model for effective evaluation of surface water quality classification using fuzzy logic concepts. The practical application of this index was demonstrated by considering both physical and inorganic chemical parameters in the study of Eber Lake, Turkey. Quality classes were combined with fuzzy logic and a single index value was obtained to represent quality classes across the breadth of parameters. As a result, a more comprehensible assessment of water quality was obtained, especially in regard to public consideration and management practices.

4.4   THE NEURAL-FUZZY APPROACH

Fuzzy neural networks are connectionist models that are trained as neural networks with their structures interpreted as a set of fuzzy rules. Models representing nonlinear input-output relationships depend on the fuzzy partition of the input-output spaces. A set of training data and a specification of rules including a preliminary definition of the corresponding membership function, $\mu_A(x)$, are required. Figure 4.7 presents a representative five-layered Takagi–Sugeno-type adaptive neuro-fuzzy inference system (ANFIS) architecture. Takagi–Sugeno models are characterized as quasi-linear in nature, resulting in a smooth transition between

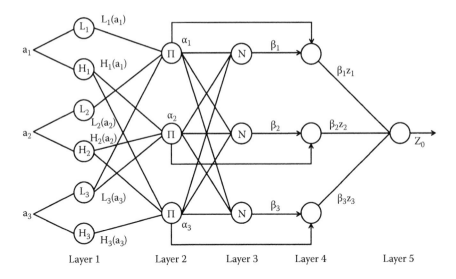

Layer 1   Layer 2   Layer 3   Layer 4   Layer 5

FIGURE 4.7 A representative five-layered Takagi–Sugeno-type adaptive neuro-fuzzy inference system (ANFIS) architecture. Detailed descriptions of the five layers are found in the main text.

linear submodels, thus allowing the separation of the identification problem into two subproblems: (1) partitioning of the state space of interest (via clustering) and (2) parameter identification of the consequent part (Lohani et al., 2006). In an ANFIS, the input and output nodes represent the input states and output response, respectively. The five layers comprising the ANFIS structure are as follows:

Layer 1: Consists of the input nodes and directly transmits the input linguistic variables to the next layer. The output is the degree to which the given input satisfies the linguistic label associated with this node.

Layer 2: All potential rules between the inputs are formulated by applying a fuzzy intersection (AND). Each node, labeled Π, computes the firing strength of the associated rule:

$$\alpha_1 = L_1\,(\alpha_1) \wedge L_2\,(\alpha_2) \wedge L_3\,(\alpha_3) \tag{4.2}$$

$$\alpha_2 = H_1\,(\alpha_1) \wedge H_2\,(\alpha_2) \wedge L_3\,(\alpha_3) \tag{4.3}$$

$$\alpha_3 = H_1\,(\alpha_1) \wedge H_2\,(\alpha_2) \wedge H_3\,(\alpha_3) \tag{4.4}$$

Layer 3: All nodes, labeled by N, indicate the normalization of firing levels of the corresponding rule:

$$\beta_1 = \frac{\alpha_1}{\alpha_1 + \alpha_2 + \alpha_3} \tag{4.5}$$

$$\beta_2 = \frac{\alpha_2}{\alpha_1 + \alpha_2 + \alpha_3} \qquad (4.6)$$

$$\beta_3 = \frac{\alpha_3}{\alpha_1 + \alpha_2 + \alpha_3} \qquad (4.7)$$

Layer 4: Outputs are the products of the normalized firing levels and individual rule outputs of the corresponding rules:

$$\beta_1 z_1 = \beta_1 HA^{-1}(\alpha_1) \qquad (4.8)$$

$$\beta_2 z_2 = \beta_2 MA^{-1}(\alpha_2) \qquad (4.9)$$

$$\beta_3 z_3 = \beta_3 SA^{-1}(\alpha_3) \qquad (4.10)$$

Layer 5: Computes the overall output as the summation of all incoming signals from Layer 4:

$$z_0 = \beta_1 z_1 + \beta_2 z_2 + \beta_3 z_3$$

The ANFIS uses a feedforward network to search for fuzzy decision rules that perform soundly on a given task. Using a specified input-output data set, an ANFIS creates a fuzzy inference system whose membership function parameters are adjusted using back-propagation alone or in combination with a least-squares method (Aqil et al., 2007). This process enables fuzzy systems to learn from the data being modeled. The hybrid learning algorithm of Jang and Gulley (1995) is routinely adopted, which consists of two alternating phases: (1) the steepest descent method, which computes error signals recursively from the output layer backward to the input nodes, and (2) the least-squares method, which finds a feasible set of consequent parameters (Aqil et al., 2007). Given fixed values of elements of antecedent parameters, the overall output is expressed as a linear combination of the consequent parameters, with the ultimate goal of minimizing the squared error. The error measure is a mathematical expression that computes the difference between the network's actual output and the desired output. ANFIS models have been widely used in both biological and environmental analyses, including, for example, the modeling of PCP-based NMDA receptor antagonists (Erdem et al., 2007), performing structural classification of proteins (Hering et al., 2003), predicting groundwater vulnerability (Dixon, 2005), analysis and prediction of flow in river basins (Aqil et al., 2007), and modeling level change in lakes and reservoirs in response to climatic variations (Yarar et al., 2009). In each case, the designed fuzzy inference systems derived reasonable actions with respect to each given rule base.

4.4.1 GENETIC ALGORITHMS IN DESIGNING FUZZY RULE-BASED SYSTEMS

The innovative combination of GAs with ANFIS models has also been explored, for example, in assessing the performance of industrial wastewater treatment processes (Chen et al., 2003). Recall that each soft computing method has advantages and disadvantages with respect to time and complexity and the amount of prior information required. When combined, this three-state analysis tool aids in deriving representative state functions for use in simulating system behavior, with the global search capability of the GA instrumental in saving rule-matching time of the inference system (Chen et al., 2003). In fact, it has been reported that fuzzy recombination allows the attainment of a reduction of nearly 40% in terms of convergence time with respect to extended line recombination operators (De Falco, 1997). This is not unexpected given that prior knowledge in the form of linguistic variables, fuzzy membership function parameters, and fuzzy rules have been shown to be easily incorporated into the genetic design process (Sivanandam and Deepa, 2008). As a result, the generation of intelligible and dependable fuzzy rules by a self-learning adaptive method is possible, while at the same time providing desirable flexibility in choosing fuzzy rules for the fuzzy logic controller (Mohammadian and Stonier, 1994; Cordon and Herrera, 2001). The controller in turn monitors the dissimilarity of the decision variables during the algorithmic process and modifies the boundary intervals appropriately to help troubleshoot and aid in finding optimal solutions to specified problems at hand.

4.5 HYBRID NEURAL NETWORK-GENETIC ALGORITHM APPROACH

Although there are numerous types of optimization algorithms for neural network structures, they do not always lead to the construction of functionally efficient networks, particularly when considering the prevention of overfitting and the reduction of the number of iterations. Given our previous coverage of genetic algorithms, and the fact that they are effective at exploring a complex space in an adaptive way, one can infer their ability to serve as intelligent neural network optimization techniques. Genetic optimization has proved valuable in the determination of efficient neural network structure through a population of individuals that evolves toward optimum solutions through genetic operators (selection, crossover, and mutation). This combination has facilitated the development of neural approaches through the generation of optimal feature selection, network connectivity weights, network architecture, and evolution of learning rules. Building upon theoretical GA concepts provided in Section 4.2, an initial population of random binary strings, each of which represented a specific neural network topology and set of training parameters, was defined. The fitness of each trained network can be calculated based on the following equation (Riveros et al., 2009):

$$\text{Fitness} = W_1 f_e(\hat{e}) + W_2 f_1(\hat{t}) \tag{4.11}$$

where $f_e(\hat{e})$ = the error between the real output and the neural network output, $f_t(\hat{t})$ = the training time of the neural network, and W_1 and W_2 = suitable values of weight. According to the fitness observed, the GA selects a new group of strings, which ultimately represents the parents of the next generation of the GA with an assigned probability of reproduction. Note that the fitness function takes into account error and the learning speed, weighting these contributions in a vigorous manner.

The strings are then subjected to the evolutionary operators defined in Section 4.2. Recall that crossover is one of the most important GA operators; it combines two chromosomes (parents) to produce a new chromosome (offspring). The crossover process occurs during evolution according to a user-definable crossover probability. The mutation process ensures that the probability of searching regions in the problem space is never zero and prevents complete loss of genetic material through reproduction and crossover. The foregoing process is repeated until the maximum number of generations is reached. Once reached, the best string that gave the maximum fitness or *MSE* is considered appropriate. The evolution process can also be terminated when there is no obvious change of best individuals after a fixed number of generations. By and large, the ANN-GA approach has shown great promise in providing powerful classification capabilities with tuning flexibility for both performance and cost-efficiency (Shanthi et al., 2009). Moreover, it has been successfully applied in numerous optimization problems, including those with a large number of continuous or discrete design variables.

As an example application, Riveros et al. (2009) described the successful use of neural networks in the on-capillary derivatization of a single-step organic reaction. More specifically, the authors empowered the use of a ANN-GA approach in the derivatization of the dipeptide D-Ala-D-Ala by phthalic anhydride (Figure 4.8) in an open-tubular format. Derivatization is mainly used for enhancement of the detection sensitivity in capillary electrophoresis (CE), for which a combination of fluorescence labeling and laser-induced fluorescence detection is favorable (Underberg and Waterval, 2010). Analysis of microreactions in an open-tubular format is one area that the small sample requirements of CE has allowed for facile study of a myriad of compounds, with a variety of both enzyme-catalyzed and organic-based reactions having been detailed (Patterson et al., 1996; Zhao and Gomez, 1998). As expected, electrophoretic-based techniques require that consideration be given to a host of parameters in order to ascertain the optimal possible conditions for selected applications. However, information processing techniques such as neural networks that provide nonlinear modeling and optimization of electrophoretic conditions, especially in regards to on-capillary derivatization, have been limited.

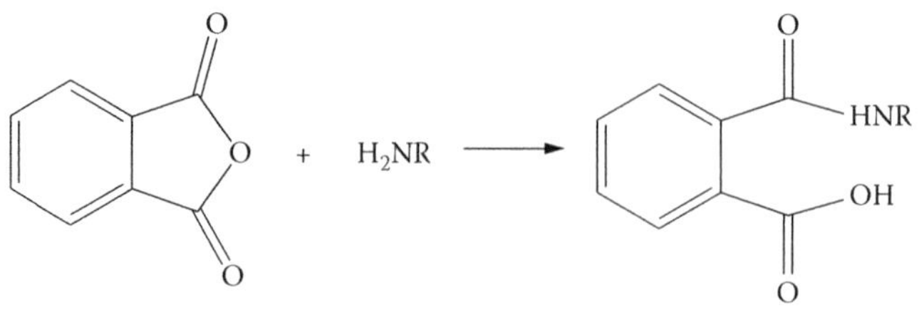

FIGURE 4.8 A general derivatization scheme of amines by phthalic anhydride.

Sixty-four on-capillary derivatizations were used as the study data set, which was randomly divided into training (46 derivatizations), test (9 derivatizations), and validation (9 derivatizations) sets. During training applications, both inputs and outputs were converted by normal transformation (mean 0, standard deviation 1). Figure 4.9 depicts the hybrid strategy of employing Levenberg–Marquardt back-propagation with a sigmoid transfer function used to search for optimal neural network architecture, and to aid in optimizing experimental conditions for maximum conversion to product. Three parameters, phthalic anhydride injection volume, time of injection, and voltage, were used as data inputs. The GA operated according to the general two-step process previously discussed: (1) initialization of the population and evaluation of each member of the initial population and (2) reproduction until a stopping condition was met. The fitness of each trained network was calculated by the use of Equation 4.11. For this study, a two-point crossover approach, one that calls for two points to be selected on the parent organism string, was used. The outlined process was repeated until the maximum number of generations was reached. Once reached, the best string that gave the maximum fitness or *MSE* was considered appropriate. For objective appreciation, synaptic weights and bias were initialized to 0.01 and 0.1, respectively. GA configuration values included a population size of 30, maximum generation of 100, a crossover probability of 0.5, and a mutation probability of 0.01.

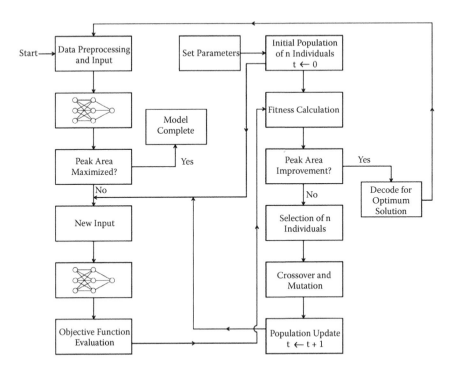

FIGURE 4.9 Schematic of the hybrid artificial neural network-genetic algorithm (ANN-GA) method for the optimization of an on-capillary dipeptide derivatization. (From Riveros et al. 2009. *Analyst* 134: 2067–2070. With permission from the Royal Society of Chemistry.)

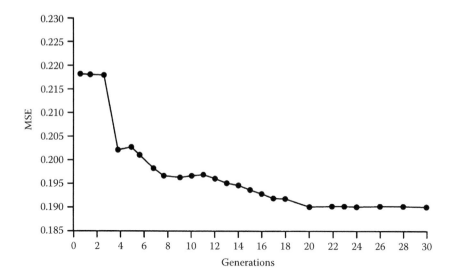

FIGURE 4.10 Plot of best mean squared error (*MSE*) versus number of generations. (From Riveros et al. 2009. *Analyst* 134: 2067–2070. With permission from the Royal Society of Chemistry.)

The optimal neural architecture (realized in 20 generations) included three hidden neurons and an *MSE* of 0.193. Figure 4.10 shows a plot of *MSE* versus generations for this specified genetic algorithm run. As justification for the optimized model, the authors ran a back-propagated neural network without the GA operator and compared this with the ANN-GA in terms of training data, with the neural model proving to be inferior to the ANN-GA approach in regard to fitting ($r^2 = 0.87$ [$n = 10$] versus $r^2 = 0.96$ [$n = 10$]) as well as predictive abilities ($r^2 = 0.90$ [$n = 10$] versus $r^2 = 0.94$ [$n = 10$]). Sensitivity analyses were performed to specify appropriate GA parameters (random number, crossover probability, mutation probability) to ensure efficiency of the solution process and global optimality of the obtained solution.

From the data patterned by the ANN-GA, the investigators generated a response surface for the two interactive factors: phthalic anhydride injection volume and voltage, which were determined by an initial 2^3-full factorial screening design. Factor significance was calculated in Analysis of Variance (ANOVA) models that were estimated and run up to their first-order interaction terms. ANOVA for a linear regression partitions the total variation of a sample into components. Validation of experimental results was accomplished by performing electrophoresis of D-Ala-D-Ala and phthalic anhydride at experimental conditions dictated by the ANN-GA model. A representative electropherogram is shown in Figure 4.11, which demonstrates that derivatization of D-Ala-D-Ala by phthalic anhydride is nearly complete. While the authors presented work valuable for optimizing on-capillary derivatization of single-step organic reactions, they stress the need to expand this methodology

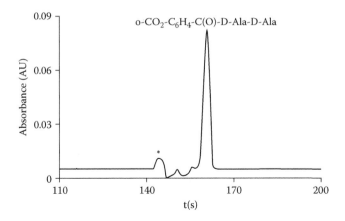

FIGURE 4.11   Representative electropherogram of the reaction product $o\text{-}CO_2\text{-}C_6H_4\text{-}D\text{-}$ Ala-D-Ala derivatized by phthalic anhydride. Reaction was carried out in 20 mM phosphate buffer (pH 9.4). The total analysis time was 6.2 min at 25 kV using a 40.0 cm (inlet to detector) coated capillary. The peak marked * is unreacted peptide. (From Riveros et al. 2009. *Analyst* 134: 2067–2070. With permission from the Royal Society of Chemistry.)

for analysis of other organic-based reaction systems, including multistep ones, and in examining other parameters associated with these reactions.

The foregoing example demonstrates the ability of the hybrid neural approach to manage experimental and computational convolution by illustrating the ability of GAs to resourcefully identify appropriate neural input parameters required for the prediction of desired characteristics. It has proved to be effective in integrated process modeling and optimization and is reportedly better than other techniques, for example, response surface methodology, particularly for complex process models (Istadi and Amin, 2006). This approach is remarkably functional for modeling networks with a multitude of inputs and outputs. Finally, we must acknowledge that as a general-purpose optimization tool, GAs will likely be applicable to and aid in other types of analyses for which evaluation functions can be derived.

4.6   CONCLUDING REMARKS

This chapter highlights the movement toward more intelligent neural network systems through the hybridization of alternative computational strategies with established neural network structural features. Although detailed attention was given to the use of evolving strategies and fuzzy inference concepts, a multitude of alternative strategies exist that help provide neural networks with effective global search capabilities, fast convergence rates, and the ability to facilitate fundamental research where multidisciplinary and multiobjective tasks are routinely present. As a consequence of this hybridization, developed neural networks are more robustly applied, generalizable, and characterized to be computationally parsimonious.

REFERENCES

Adly, A.A., and Abd-El-Hafiz, S.K. 2008. Efficient modeling of vector hysteresis using fuzzy inference systems. *Physica B* 403: 3812–3818.

Adroer, M., Alsina, A., Aumatell, J., and Poch, M. 1999. Wastewater neutralization control based on fuzzy logic: Experimental results. *Industrial and Engineering Chemistry Research* 38: 2709–2719.

Aqil, M., Kita, I., Yano, A., and Nishiyama, S., 2007. A comparative study of artificial neural networks and neuro-fuzzy in continuous modeling of the daily and hourly behaviour of runoff. *Journal of Hydrology* 337: 22–34.

Astel, A. 2007. Chemometrics based on fuzzy logic principles in environmental studies. *Talanta* 72: 1–12.

Blanco, A., Delgado, M., and Pegalajar, M.C. 2001. A real-coded genetic algorithm for training recurrent neural networks. *Neural Networks* 14: 93–105.

Brubaker, D.I. 1992. Fuzzy-logic basics: Intuitive rules replace complex math. *Engineering Design* June 18, 1992: 111–118.

Chang, C.S., Xu, D.Y., and Quek, H.B. 1999. Pareto-optimal set based multiobjective tuning of fuzzy automatic train operation for mass transit system. *IEE Proceedings of Electric Power Applications* 146: 575–583.

Chen, W.C., Chang, N.-B., and Chen, J.-C. 2003. Rough set-based hybrid fuzzy-neural controller design for industrial wastewater treatment. *Water Research* 37: 95–107.

Choy, W.Y., and Sanctuary, B.C. 1998. Using genetic algorithms with *a priori* knowledge for quantitative NMR signal analysis. *Journal of Chemical Information and Computer Sciences* 38: 685–690.

Cordon, O., and Herrera, F. 2001. Hybridizing genetic algorithms with sharing scheme and evolution strategies for designing approximate fuzzy rule-based systems. *Fuzzy Sets and Systems* 118: 235–255.

De Falco, I. 1997. Nonlinear system identification by means of evolutionary optimised neural networks, In D. Quagliarella, J. Périaux, C. Poloni, and G. Winter (Eds.), *Genetic Algorithms and Evolution Strategies in Engineering and Computer Science*. John Wiley & Sons, West Sussex, England, pp. 367–390.

Dixon, B. 2005. Applicability of neuro-fuzzy techniques in predicting ground-water vulnerability: A GIS-based sensitivity analysis. *Journal of Hydrology* 309: 17–38.

Dubois, D., and Prade, H. 1999. Fuzzy sets in approximate reasoning, Part 1: Inference with possibility distributions. *Fuzzy Sets and Systems* 100: 73–132.

Erdem, B., Arzu, S., Murat, A., Ferda Nur, A., and Adeboye, A. 2007. Adaptive neuro-fuzzy inference system (ANFIS): A new approach to predictive modeling in QSAR applications : A study of neuro-fuzzy modeling of PCP-based NMDA receptor antagonists. *Bioorganic & Medicinal Chemistry* 15: 4265–4282.

Gao, F., Li, M., Wang, F., Wang, B., and Yue, P. 1999. Genetic algorithms and evolutionary programming hybrid strategy for structure and weight learning for multilayer feedforward neural networks. *Industrial Engineering and Chemical Research* 38: 4330–4336.

Goldberg, D.E. 1989. *Genetic Algorithms in Search, Optimization and Machine Learning*. Addison-Wesley 1: Reading, MA.

Goldberg, D.E. 1991. Real-coded genetic algorithms. Virtual alphabets, and blocking. *Complex Systems* 5: 139–167.

Gottwald, S. 2005. Mathematical aspects of fuzzy sets and fuzzy logic. Some reflections after 40 years. *Fuzzy Sets and Systems* 156: 357–364.

Hering, J.A., Innocent, P.R., and Haris, P.I. 2003. Neuro-fuzzy structural classification of proteins for improved protein secondary structure prediction. *Proteomics* 3: 1464–1475.

Holland, J.H. 1975. *Adaptation in Natural Selection and Artificial Systems*. University of Michigan Press: Ann Arbor, MI, USA.

Hrstka, O., and Kučerová, A. 2004. Improvement of real coded genetic algorithms based on differential operators preventing premature convergence. *Advances in Engineering Software* 35: 237–246.

Icaga, Y. 2007. Fuzzy evaluation of water quality classification. *Ecological Indicators* 7: 710–718.

Istadi, I., and Amin, N.A.S. 2006. Hybrid artificial neural network-genetic algorithm technique for modeling and optimization of plasma reactor. *Industrial and Engineering Chemistry Research* 45: 6655–6664.

Jang, J.-S., and Gulley, N. 1995. *The fuzzy logic toolbox for use with MATLAB*. The Mathworks, Inc., Natick, MA.

Kellegöz, T., Toklu, B., and Wilson, J. 2008. Comparing efficiencies of genetic crossover operators for one machine total weighted tardiness problem. *Applied Mathematics and Computation* 199: 590–598.

Kita, H. 2001. A comparison study of self-adaptation in evolution strategies and real-coded genetic algorithms. *Evolutionary Computation Journal* 9:223–241.

Klir, G.J., and Smith, R.M. 2001. On measuring uncertainty and uncertainty-based information: Recent development. *Annals of Mathematics and Artificial Intelligence* 32: 5–33.

Konak, A., Coit, D.W., and Smith, A.E. 2006. Multi-objective optimization using genetic algorithms: A tutorial. *Reliability Engineering and System Safety* 91: 992–1007.

Kooistra, L., Huijbregts, M.A.J., Ragas, A.M.J., Wehrens, R., and Leuven, R.S.E.W. 2005. Spatial Variability and Uncertainty in ecological risk assessment: A case study on the potential risk of cadmium for the little owl in a Dutch river flood plain. *Environmental Science and Technology* 39: 2177–2187.

Kosko, B. 1994. Fuzzy systems as universal approximators. *IEEE Transactions on Computers* 43: 1329–1333.

Koza, J.R. 1992. *Genetic Programming: On the Programming of Computers by Means of Natural Selection*. MIT Press: Cambridge, MA.

Lohani, A.K., Goel, N.K., and Bhatia, K.K.S. 2006. Takagi-Sugeno fuzzy inference system for modeling stage-discharge relationship. *Journal of Hydrology* 331: 146–160.

Mamdani, E.H. 1976. Application of fuzzy logic to approximate reasoning using linguistic synthesis. *Proceedings of the Sixth International Symposium on Multiple-Valued Logic*. IEEE Computer Society Press: Los Alamitos, CA.

Mitaim, S., and Kosko, B. 2001. The shape of fuzzy sets in adaptive function approximation. *IEEE Transactions on Fuzzy Systems* 9: 637–656.

Mitchell, M. 1998. *An Introduction of Genetic Algorithms*. MIT Press: Cambridge, MA.

Mohammadian, M., and Stonier, R.J. 1994. Generating fuzzy rules by genetic algorithms. *Proceedings of 3rd IEEE International Workshop on Robot and Human Communication*. Nagoya, pp. 362–367.

Nurcahyo, G.W., Shamsuddin, S.M., and Alias, R.A. 2003. Selection of defuzzification method to obtain crisp value for representing uncertain data in a modified sweep algorithm. *Journal of Computer Science and Technology* 3: 22–28.

Patterson, D.H., Harmon, B.J., and Regnier, F.E. 1996. Dynamic modeling of electrophoretically mediated microanalysis. *Journal of Chromatography A* 732: 119–132.

Reid, D.J. 1996. Genetic algorithms in constrained optimization. *Mathematical and Computer Modelling* 23: 87–111.

Riveros, T.A., Hanrahan, G., Muliadi, S., Arceo, J., and Gomez, F.A. (2009). On-capillary derivatization using a hybrid artificial neural network-genetic algorithm approach. *Analyst* 134: 2067–2070.

Rothlauf, F. 2006. *Representations for Genetic and Evolutionary Algorithms*. Springer: New York.

Schmidt, M., and Lipson, H. 2007. *Comparison of Tree and Graph Encodings as Function of Problem Complexity.* Genetic and Evolutionary Computation Conference (GECCO) '07, June 25–29, 2007. London, England.

Schwefel, H.-P., and Bäck, T. 1997. Artificial evolution: How and why? In D. Quagliarella, J. Périaux, C. Poloni, and G. Winter (Eds.), *Genetic Algorithms and Evolution Strategies in Engineering and Computer Science*, John Wiley & Sons, West Sussex, England, pp. 1–19.

Shanthi, D., Sahoo, G., and Saravanan, N. 2009. Evolving connection weights of artificial neural networks using genetic algorithm with application to the prediction of stroke disease. *International Journal of Soft Computing* 4: 95–102.

Silvert, W. 2000. Fuzzy indices of environmental conditions. *Ecological Modelling* 130: 111–119.

Sivanandam, S.N., and Deepa, S.N. 2008. *Introduction to Genetic Algorithms.* Springer: Heidelberg, Germany.

Sugeno, M. 1985. *Industrial Applications of Fuzzy Control.* Elsevier, Amsterdam.

Syswerda, G. 1989. Uniform crossover in genetic algorithms. In J.D. Schafer (Ed.), *Proceedings of the 3rd International Conference on Genetic Algorithms.* Morgan Kaufmann: San Francisco, pp. 2–9.

Underberg, W.J.M., and Waterval, J.C.M. 2010. Derivatization trends in capillary electrophoresis: An update. *Electrophoresis* 23: 3922–3933.

Uricchio, V.F., Giordano, R., and Lopez, N. 2004. A fuzzy knowledge-based decision support system for groundwater pollution risk evaluation. *Journal of Environmental Management* 73: 189–197.

Yang, S. 2002. Adaptive crossover in genetic algorithms using statistics mechanism. In R. Standish, H.A. Abbass, and M.A. Bedau, (Eds.), *Artificial Life VIII.* MIT Press: Cambridge, pp. 182–185.

Yarar, A., Onucyıldız, M., and Copty, N.K. 2009. Modelling level change in lakes using neurofuzzy and artificial neural networks. *Journal of Hydrology* 365: 329–334.

Zadeh, L.A. 1965. Fuzzy sets. *Information and Control* 8: 338–353.

Zadeh, L.A. 1978. Fuzzy sets as a basis for a theory of possibility. *Fuzzy Sets and Systems* 1: 3–28.

Zhao, D.S., and Gomez, F.A. 1998. Double enzyme-catalyzed microreactors using capillary electrophoresis. *Electrophoresis* 19: 420–426.

Zhao, J., and Bose, B.K. 2002. Evaluation of membership function for fuzzy logic controlled induction motor drive. *Proceeding of the IEEE 28th Annual Conference of the Industrial Electronics Society* (IECON'02). Sevilla, Spain, pp. 229–234.

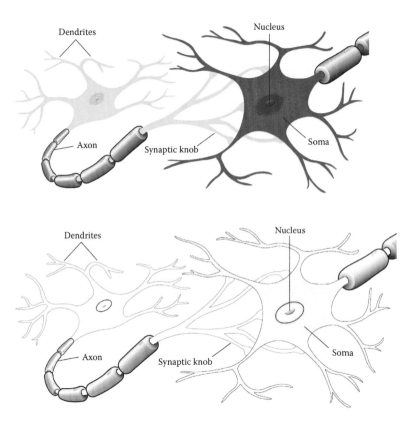

FIGURE 1.2 Biological neurons organized in a connected network, both receiving and sending impulses. Four main regions comprise a neuron's structure: the soma (cell body), dendrites, axons, and synaptic knobs. (From Hanrahan, G. 2010. *Analytical Chemistry*, 82: 4307–4313. With permission from the American Chemical Society.)

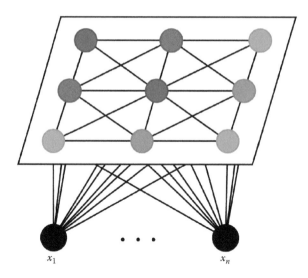

FIGURE 2.9 A Kohonen's self-organizing map (SOM) displaying a feedforward structure with a single computational layer arranged in rows and columns. For the input layer, each node is a vector representing N terms. Each output node is a vector of N weights. Upon visual inspection, colors are clustered into well-defined regions, with regions of similar properties typically found adjoining each other. (From Hanrahan, G. 2010. *Analytical Chemistry*, 82: 4307–4313. With permission from the American Chemical Society.)

Parent 1		Parent 2	
0100101	101011001	1100101	001001010
Offspring 1		**Offspring 2**	
0100101	001001010	1100101	101011001

FIGURE 4.3 Illustration of the single-point crossover process. As depicted, the two parent chromosomes are cut once at corresponding points and the sections after the cuts swapped with a crossover point selected randomly along the length of the mated strings. Two offspring are then produced.

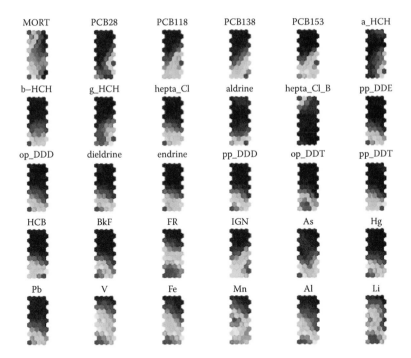

FIGURE 6.4 Representative subset of SOMs for 30 of the 44 parameters tested in chronic toxicity mode. Make special note of the values from the mortality SOM, where a limited number of sites revealed high mortality. (From Tsakovski et al. 2009. *Analytica Chimica Acta* 631: 142–152. With permission from Elsevier.)

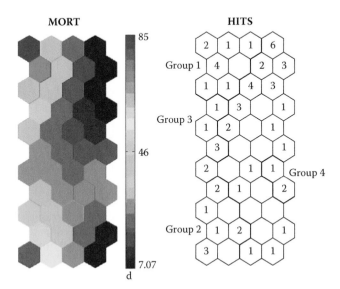

FIGURE 6.5 The vicinities of four groups of sites that revealed high levels of mortality. (From Tsakovski et al. 2009. *Analytica Chimica Acta* 631: 142–152. With permission from Elsevier.)

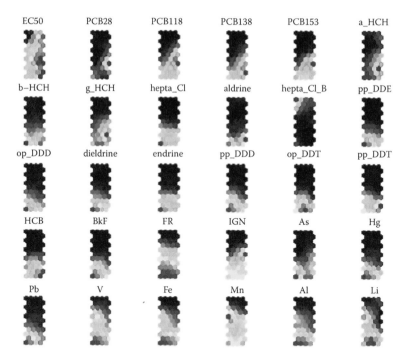

FIGURE 6.6 Representative subset of SOMs for 30 of the 44 parameters tested in acute toxicity mode. The major indicator observed was EC$_{50}$, whose values were indicative of the acute toxicity of the collected samples. (From Tsakovski et al. 2009. *Analytica Chimica Acta* 631: 142–152. With permission from Elsevier.)

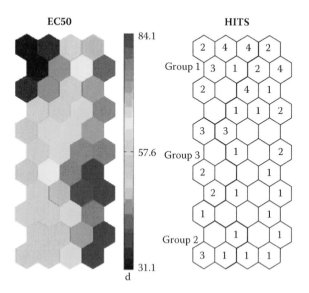

FIGURE 6.7 Highlight of three distinct groups of similar acute toxicity and object distribution. (From Tsakovski et al. 2009. *Analytica Chimica Acta* 631: 142–152. With permission from Elsevier.)

5 Applications in Biological and Biomedical Analysis

5.1 INTRODUCTION

Neural network methods are recognized as holding great promise for advanced understanding of biological systems, associated phenomena, and related processes. For example, the ability of neural networks to learn arbitrary complex functions from large amounts of data without the need for predetermined models makes them an ideal tool for protein structure prediction (Wood and Hirst, 2005). Others have utilized the powerful classification capabilities of neural networks for the analysis and unsupervised classification of electron microscopy images of biological macromolecules (Pascual-Montano et al., 2001). Moreover, they can aid in acquiring accurate knowledge of quantitative structure–activity relationships (e.g., Jalali-Heravi et al., 2008; Lara et al., 2008), a process by which chemical structure is quantitatively correlated with biological activity (or chemical reactivity). Regardless of the application area, neural networks can offer sophisticated modeling tools when faced with a profusion of data, even if there is an inadequate understanding of the biological system under investigation. Yet, despite the increased intricacy in model representations, in practice it has been demonstrated that anticipated improvements have frequently not lived up to expectations. Having said that, the applications presented in this chapter provide the reader with encouraging detail in regard to the use of neural networks in contemporary biological research efforts, and comment directly on ways in which developed models can be enhanced. Model framework, development, validation, and application concepts covered comprehensively in previous chapters will provide context for the coverage of these investigations.

5.2 APPLICATIONS

A variety of innovative neural network applications are summarized in this chapter, with a detailed list of examples provided in Table 5.1. Included in this table is fundamental information regarding application areas, model descriptions, key findings, and overall significance of the work. Representative examples from this table are expanded upon in the main text. This list is by no means comprehensive, but it does provide a representative view of the flexibility of neural network modeling and its widespread use in the biological and biomedical sciences. Readers are encouraged to explore the literature in greater detail, particularly in regard to the evolution of this field, and critically assess the influence of such developments on forthcoming investigative research and quantitative assessment activities.

TABLE 5.1

Selected Neural Network Model Applications in Modern Biological and Biomedical Analysis Efforts

Analyte/Application Area	Model Description	Key Findings/Significance	Reference
Optimized separation of neuroprotective peptides	Experimental design (ED) approach for suitable input/output data sources for feedforward ANN training	The combined ED-ANN approach was found to be effective in optimizing the reverse-phase high-performance liquid chromatography (RPLC) separation of peptide mixtures	Novotná et al. (2005)
Simultaneous determination of ofloxacin, norfloxacin, and ciprofloxacin	Radial basis function–artificial neural network (RBF-ANN)	The RBF-ANN calibration model produced the most satisfactory figures of merit and was subsequently used for prediction of the antibiotics of bird feedstuff and eye drops	Ni et al. (2006)
Improved peptide elution time prediction in reverse-phase liquid chromatography	Feedforward neural network. A genetic algorithm (GA) was used for the normalization of any potential variability of the training retention time data sets	Approximately 346,000 peptides were used for the development of a peptide retention time predictor. The model demonstrated good elution time precision and was able to distinguish among isomeric peptides based on the inclusion of peptide sequence information	Petritis et al. (2006)
Analysis of bacterial bioreporter response obtained with fluorescence flow cytometry	Single-layer perceptron artificial neural network (SLP-ANN) based on sequential parameter estimation	Successful analysis of flow cytometric data for bioreceptor response for both arsenic biosensing and HBP (strain Str2-HBP) applications	Busam et al. (2007)
Identification of new inhibitors of P-glycoprotein (P-gp)	Self-organizing maps (SOMs)	Self-organizing maps were effectively trained to separate high- and low-active propafenone-type inhibitors of P-gp	Kaiser et al. (2007)

Prediction of peptide separation in strong anion exchange (SAX) chromatography	MLP used as a pattern classifier. A genetic algorithm (GA) was employed to train the neural network	In this study, an average classification success rate of 84% in predicting peptide separation on a SAX column using six features to describe each peptide. Out of the six features, sequence index, charge, molecular weight, and sequence length make significant contributions to the prediction	Oh et al. (2007)
Modeling and optimization of fermentation factors and evaluation for alkaline protease production	Hybrid genetic algorithm-artificial neural network (GA-ANN) approach	GA-ANN was successfully used in the optimization of fermentation conditions (incubation temperature, medium pH, inoculum level, medium volume, and carbon and nitrogen sources) to enhance the alkaline protease production by *Bacillus circulans*.	Rao et al. (2008)
Quantitative structure-activity relationship (QSAR) study of heparanase inhibitor activity	Comparison of Levenberg–Marquardt (L-M), back-propagation (BP), and conjugate gradient (CG) algorithms	A 4-3-1 L-M neural network model using leave-one-out (LOO), leave-multiple out (LMO) cross validation, and Y-randomization was successful in studying heparanase inhibitors	Jalali-Heravi et al. (2008)
Analysis of the affinity of inhibitors for HIV-1 protease	Genetic algorithm optimized fuzzy neural network (GA-FNN) employing tenfold cross-validation	A fuzzy neural network (FNN) was trained on a data set of 177 HIV-1 protease ligands with experimentally measured IC_{50} values. A genetic algorithm was used to optimize the architecture of the FNN used to predict biological activity of HIV-1 protease inhibitors	Fabry-Asztalos et al. (2008)
Prediction of antigenic activity in the hepatitis C virus NS3 protein	Feedforward neural network employing back propagation with momentum learning algorithm	The neural network model was capable of predicting the antigenic properties of HCV NS3 proteins from sequence information alone. This allowed an accurate representation of quantitative structure-activity relationship (QSAR) of the HCV NS3 conformational antigenic epitope	Lara et al. (2008)
Design of small peptide antibiotics effective against antibiotic-resistant bacteria	QSAR methodology combined with a feedforward neural network	Two random 9-amino-acid peptide libraries were created with the resultant data fed into a feedforward neural network. As a result, quantitative models of antibiotic activity were created	Cherkasov et al. (2008)

(Continued)

TABLE 5.1 (CONTINUED)
Selected Neural Network Model Applications in Modern Biological and Biomedical Analysis Efforts

Analyte/Application Area	Model Description	Key Findings/Significance	Reference
Optimization of reaction conditions for the conversion of nicotinamide adenine dinucleotide (NAD) to reduced form (NADH)	Experimental design (ED) approach for suitable input/output data sources for feedforward, back-propagated network training	A full factorial experimental design examining the factors' voltage (V), enzyme concentration (E), and mixing time of reaction (M) was utilized as input-output data sources for suitable network training for prediction purposes. This approach proved successful in predicting optimal conversion in a reduced number of experiments	Riveros et al. (2009a)
Optimization of on-capillary dipeptide (D-Ala-D-Ala) derivatization	Hybrid genetic algorithm–artificial neural network (GA-ANN) approach	Results obtained from the hybrid approach proved superior to a neural network model without the GA operator in terms of training data and predictive ability. The model developed is a potential tool for the analysis of other organic-based reaction systems	Riveros et al. (2009b)
Prediction of antifungal activity of pyridine derivatives against *Candida albicans*	Feedforward neural network employing the Broyden–Fletcher–Goldfarb–Shanno (BFGS) learning algorithm	The neural network model proved effective with respect to prediction of antimicrobial potency of new pyridine derivatives based on their structural descriptors generated by calculation chemistry	Buciński et al. (2009)
Classification of the life-cycle stages of the malaria parasite	Multilayer perceptron feedforward neural network	Efficient training of the neural network model allowed detailed examination of synchrotron Fourier transform infrared (FT-IR) spectra, with discrimination between infected cells and control cells possible	Webster et al. (2009)
Optimization of HPLC gradient separations as applied to the analysis of benzodiazepines in postmortem samples	Multilayer perceptron feedforward neural network in combination with experimental design	Neural networks were used in conjunction with experimental design to efficiently optimize a gradient HPLC separation of nine benzodiazepines. The authors report a more flexible and convenient means for optimizing gradient elution separations than was previously reported	Webb et al. (2009)

Prediction of the isoforms' specificity of cytochrome P450 substrates	Counter-propagation neural networks (CPG-NN)	The CPG-NN approach proved valuable as a graphical visualization tool for the prediction of the isoform specificity of cytochrome P450 substrates	Michielan et al. (2009)
Variable selection in the application area of metabolic profiling	Self-organizing maps (SOMs)	Provides details on the extension of the SOM discrimination index (SOMDI) for classification and determination of potentially discriminatory variables. Methods are illustrated in the area of metabolic profiling consisting of an NMR data set of 96 saliva samples	Wongravee et al. (2010)
Quantitative analysis of mebendazole polymorphs A-C	Feedforward, back-propagated neural network after PCA compression	A method based on diffuse reflectance FTIR spectroscopy (DRIFTS) and neural network modeling with PCA input space compression allowed the simultaneous quantitative analysis of mebendazole polymorphs A-C in power mixtures	Kachrimanis et al. (2010)

5.2.1 Enzymatic Activity

Enzymes are highly specific and efficient organic catalysts, with activity highly dependent on numerous factors, including temperature, pH, and salt concentration (Tang et al., 2009). Although electrophoretically mediated microanalysis (EMMA) and related capillary electrophoresis (CE) methods have been widely applied to measuring enzyme activities and other parameters (e.g., Zhang et al., 2002; Carlucci et al., 2003), little work has been devoted to optimizing the experimental conditions for these techniques. CE comprises a family of techniques including capillary zone electrophoresis, capillary gel electrophoresis, isoelectric focusing, micellar electrokinetic capillary chromatography, etc. Such techniques employ narrow-bore (e.g., 20–200 μm i.d.) capillaries to achieve high efficiency separations for the laboratory analysis of biological materials and are unparalleled experimental tools for examining interactions in biologically relevant media (Hanrahan and Gomez, 2010). A generalized experimental setup for CE is presented in Figure 5.1. As shown, the instrumental configuration is relatively simple and includes a narrow-bore capillary, a high-voltage power supply, two buffer reservoirs, a sample introduction device, and a selected detection scheme, typically UV-visible or laser-induced fluorescence (LIF). In EMMA, differential electrophoretic mobility is used to merge distinct zones of analyte and analytical reagents under the influence of an electric field (Zhang et al., 2002; Burke and Reginer, 2003). The reaction is then allowed to proceed within the region of reagent overlap either in the presence or absence of an applied potential, with the resultant product being transported to the detector under the influence of an electric field.

Previously, Kwak et al. (1999) used a univariate approach to optimizing experimental conditions for EMMA, more specifically, the optimization of reaction conditions for the conversion of nicotinamide adenine dinucleotide (NAD) to nicotinamide

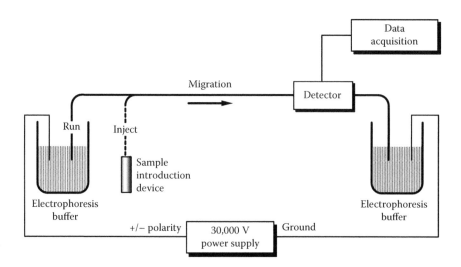

FIGURE 5.1 A generalized capillary electrophoresis experimental setup (From Hanrahan and Gomez. 2010. *Chemometric Methods in Capillary Electrophoresis*. John Wiley & Sons, Hoboken, N.J. With permission from John Wiley & Sons, Inc.).

adenine dinucleotide, reduced form (NADH), by glucose-6-phosphate dehydrogenase (G6PDH, EC 1.1.1.49) in the conversion of glucose-6-phosphate (G6P) to 6-phosphogluconate. More recently, our group made use of response surface methodology (RSM) in the form of a Box-Behnken design using the same G6PDH model system (Montes et al., 2008). The Box-Behnken design is considered an efficient option in RSM and an ideal alternative to central composite designs. It has three levels per factor, but avoids the corners of the space, and fills in the combinations of center and extreme levels. It combines a fractional factorial with incomplete block designs in such a way as to avoid the extreme vertices and to present an approximately rotatable design with only three levels per factor (Hanrahan et al., 2008). In this study, the product distribution—product/(substrate + product)—of the reaction was predicted, with results in good agreement (7.1% discrepancy difference) with the experimental data. The use of chemometric RSM provided a direct relationship between electrophoretic conditions and product distribution of the microscale reactions in CE and has provided scientists with a new and versatile approach to optimizing enzymatic experimental conditions. There have also been a variety of additional studies incorporating advanced computational techniques in CE, including, for example, optimizing the separation of two or more components via neural networks (e.g., Zhang et al., 2005). In this selected literature reference, the investigators applied an MLP neural network based on genetic input selection for quantification of overlapping peaks in micellar electrokinetic capillary chromatography (MECC).

The aim of a 2009 study by our group was to demonstrate the use of natural computing, in particular neural networks, in improving prediction capabilities and enzyme conversion in EMMA. A full factorial experimental design examining the factors voltage (V), enzyme concentration (E), and mixing time of reaction (M) was utilized as input data sources for suitable network training for prediction purposes. This type of screening design is vital in determining initial factor significance for subsequent optimization. It is especially important in CE method development, where the most influential factors, their ranges, and interactions are not necessarily known. This combined approach was patterned after the seminal work of Havel and colleagues (Havel et al., 1998), whose use of experimental design techniques for proper neural network input was significant in defining future studies. To evaluate the influence of mixing time, voltage, and enzyme concentration on the percentage conversion of NAD to NADH by glucose-6-phosphate dehydrogenase, we employed a 2^3 factorial design. The eight randomized runs and acquired data obtained are highlighted in Table 5.2. Statistical analysis of the model equations revealed r^2 (0.93) and adjusted r^2 (0.91) values. An examination of Prob>F from the effect test results revealed that enzyme concentration had the greatest single effect (Prob>F = <0.001). Prob>F is the significance probability for the F-ratio, which states that if the null hypothesis is true, a larger F-statistic would only occur due to random error. Significant probabilities of 0.05 or less are considered evidence of a significant regression factor in the model. Additionally, a significant interactive effect (Prob>F = 0.031) between mixing time and voltage was revealed.

In order to optimize the conversion of NAD to NADH by glucose-6-phosphate dehydrogenase, an optimal 3:4:1 feedforward neural network structure (Figure 5.2) generated using information obtained from the 2^3 factorial screening design was

TABLE 5.2
Results from the 2^3 Factorial Design in Riveros et al. (2009a)[a]

Experiment	Mixing Time (min)	Voltage (kV)	Enzyme Concentration (mg/mL)	Mean Percentage Conversion (Experimental, $n = 3$)	R.S.D. (%) (Experimental, $n = 3$)	Percentage Conversion (Predicted)	Percentage Difference
1	0.2	1	1	8.68	8.21	7.99	7.9
2	0.2	1	7	12.6	4.91	13.3	5.6
3	0.2	25	7	10.5	5.15	12.1	15.2
4	0.2	25	1	4.99	13.6	5.17	3.6
5	1.4	25	7	11.2	1.17	10.5	6.3
6	1.4	25	1	8.93	7.88	9.21	3.1
7	1.4	1	1	17.8	2.99	18.5	3.9
8	1.4	1	7	36.4	6.70	34.7	4.7

[a] Riveros et al. 2009a. *Electrophoresis* 30: 2385–2389. With permission from Wiley-VCH.

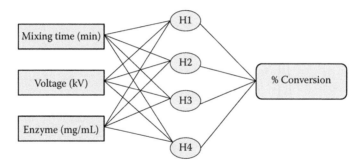

FIGURE 5.2   An optimal 3:4:1 feedforward network structure employed in Riveros et al. (2009a). (With permission from Wiley-VCH.)

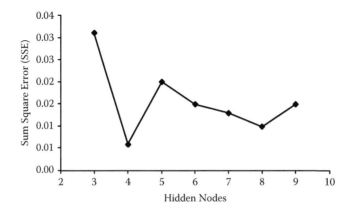

FIGURE 5.3   Sum square error (SSE) values versus the number of hidden nodes of input data. (From Riveros et al. 2009a. *Electrophoresis* 30: 2385–2389. With permission from Wiley-VCH.)

developed. Refer to Figure 5.3 for visualization of optimal hidden node determination. Here, the number of nodes were varied from 3 to 9 and plotted against the sum square error (*SSE*). As shown, four hidden nodes resulted in the lowest *SSE* with no further improvement upon increasing the hidden node number. To select the optimum number of iterations, examination of the mean square error (*MSE*) of the training set and testing set versus learning iterations was performed. Here, the number of iterations was stopped at 7,500, a value where the error for the data set ceased to decrease. Upon adequate network structure determination (3:4:1) and model development, a data subset in the range selected in the experimental design was created with the neural network used for prediction purposes, ultimately searching for optimized percentage conversion. From the data patterned by the network, a contour profile function was used to construct a response surface for the two interactive factors (mixing time and voltage). This interactive profiling facility was employed for optimizing the response surface graphically with optimum predicted values of mixing time = 1.41 min, voltage = 1.2 kV, and with enzyme concentration

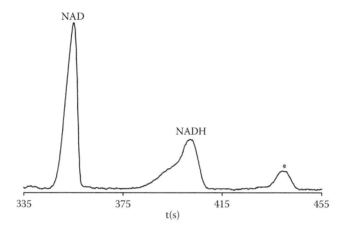

FIGURE 5.4   Representative electropherogram showing the separation of NAD and NADH after reaction with G6DPH in 50 mM Borate, 200 μM G6P buffer (pH 9.44). The total analysis time was 8.45 min at 1.0 kV (92.8 μA) using a 40.0 cm (inlet to detector) coated capillary. The peak marked * is an impurity. (From Riveros et al. 2009a. *Electrophoresis* 30: 2385–2389. With permission from Wiley-VCH.)

held constant at 1.00 mg mL^{-1}. These conditions resulted in a predicted conversion of 42.5%.

To make evident the predictive ability of the developed model, a series of three repeated experiments using the modeled optimal conditions listed earlier were carried out. A representative electropherogram from replicate number two is shown in Figure 5.4. While the peak for NAD is sharp in the electropherogram, the peak for NADH is expansive and tails in the front end. On continued electrophoresis, the concentration of NAD in the plug that is overlapped with the plug of enzyme reached its maximum, resulting in the optimal conversion rate to product (greatest height of the NADH peak). Stacking of the product plug occurs on continued electrophoresis, resulting in the characteristic peak shape at the end of the overlap of the two plug zones.

Realizing that neural network modeling capabilities do not always result in good generalizability, we ran a general linear model (GLM), ostensibly running a neural network without a hidden layer, and compared this to our hidden layer model in terms of training data. Examination was made with respect to the corrected c-index (concordance index), where a c-index of 1 indicates a "perfect" model and a c-index of 0.5 indicates a model that cannot predict any better than an indiscriminate model. The mean c-index for the hidden layer model was 0.8 ± 0.1, whereas the GLM registered 0.6 ± 0.1. Additionally, we employed the Akaike Information Criteria (AIC) for further assessment of neural network model generalizability. The AIC is a method of choosing a model from a given set of models. The chosen model is the one that minimizes the Kullback–Leibler distance between the model and the truth. In essence, it is based on information theory, but a heuristic way to think about it is as a criterion that seeks a model that has a good fit to the truth but with few parameters (Burnham and Anderson, 2004). In this study, the AIC was used to compare the two models with the same training set data. At this point, we assessed the related error term (the model that had the lowest AIC was considered to be the best). This proved valuable in our selection of the network hidden layer model.

There were systematic negative relative differences displayed between the predicted model and experimental results. A likely criticism comes in the form of the "Black Box" discussion, where models are considered applicable only within a given system space. We acknowledge that our training data set was not overly large and likely resulted in predictions slightly away from the range of the training data. We have, nonetheless, presented a representative subset (in statistical terms) through the incorporation of systematic experimental design procedures. More noteworthy, our neural network model allowed extrapolation and prediction beyond our initial range of chosen factors in the factorial design. As a result, percentage conversion (experimental) increased substantially from the factorial design, and also when compared to our previous use of a Box-Behnken response surface model alone in a similar EMMA study (Montes et al., 2008). The input patterns required for neural network training in this work necessitated the use of merely 8 experimental runs through a full factorial design. This is compared to our previous work using RSM alone, which required a total of 15 experimental runs to acquire appropriate model predicted values. Moreover, the use of a neural network approach reduced the amount of NAD required in the optimization studies from 500 to 130 picomoles.

5.2.2 QUANTITATIVE STRUCTURE–ACTIVITY RELATIONSHIP (QSAR)

Quantitative structure–activity relationship (QSAR) studies endeavor to associate chemical structure with activity using dedicated statistical and computational approaches, with the assumption that correlations exist between physicochemical properties and molecular structure (Livingstone, 2000; Guha et al., 2005; Jalali-Heravi and Asadollahi-Baboli, 2009). QSAR and other related approaches have attracted broad scientific interest, chiefly in the pharmaceutical industry for drug discovery and in toxicology and environmental science for risk assessment. In addition to advancing our fundamental knowledge of QSAR, these efforts have encouraged their application in a wider range of disciplines, including routine biological and chemical analysis. QSAR has also matured significantly over the last few decades, accounting for more highly developed descriptors, models, and selection of substituents. When physicochemical properties or structures are expressed numerically, investigators can fashion a defined mathematical relationship. For coding purposes, a number of features or molecular descriptors are calculated. Descriptors are parameters calculated from molecular structure. They can also be measured by assorted physicochemical methods. Realizing that molecular descriptors can lack structural interpretation ability, investigators will frequently employ fuzzy logic, genetic algorithms, and neural network approaches to fully explore the experimental domain. An advantage of neural network techniques over traditional regression analysis methods is their inherent ability to incorporate nonlinear relationships among chemical structures and physicochemical properties of interest.

In a representative study, a computational model developed by Lara et al. (2008) defined QSAR for a major conformational antigenic epitope of the hepatitis C virus (HCV) nonstructural protein 3 (NS3). It has been shown that immunoreactive forms of HCV antigens can be used for diagnostic assays involving characterization of antigenic determinants derived from different HCV strains (Lin et al., 2005). The same authors, among others (e.g., Khudyakov et al., 1995), showed that the HCV NS3 protein contained conformation-dependent immunodominant B cell epitopes, with one of the antigenic regions having the ability to be modeled with recombinant proteins of 103 amino acids long. Using this as a base of experimentation, Lara and colleagues applied QSAR analysis to investigate structural parameters that quantitatively define immunoreactivity in this HCV NS3 conformational antigenic region. The data set consisted of 12 HCV NS3 protein variants encompassing the amino acid positions 331–433 (HCV NS3 helicase domain) or positions 1357–1459 (HCV polyprotein). Variants were tested against 115 anti-HCV positive serum samples. Of the 115 samples, 107 were included in the neural network model training set described in the following text.

A fully connected feedforward neural network trained using error propagation with a momentum learning algorithm was employed. Error back-propagation (also routinely termed the generalized delta rule) was used as the cost function for updating the weights and minimization of error. Recall from our previous discussion that the generalized delta rule, developed by Rumelhart and colleagues, is similar to the delta rule proposed by Widrow and Hoff and one of the most often-used supervised learning algorithms in feedforward, multilayered networks. Here, the adjustment of weights leading to the hidden layer neurons occurs (in addition to the typical adjustments to the weights leading to the output neurons). In effect, using the generalized delta rule to fine-tune the weights leading to the hidden units is considered back-propagating the error adjustment. In this study, results of a stepwise optimization approach revealed the optimal size of the neural network architecture (159 hidden units) and a 1,500 iteration training cycle. The learning rate was set to 0.1 and the momentum to 0.3. Upon optimization, the neural network was trained to map a string of real numbers representing amino acid physiochemical properties onto 107 real-valued output neurons corresponding to the enzyme immunoassay (EIA) Signal/Cutoff (S/Co) values. Note that proper sequence-transforming schemes for protein sequence representation was performed to ensure quality neural network performance. See Figure 5.5 for the generated HCV NS3 sequences. In addition, relevant molecular modeling studies were carried out for position mapping. These processes are described in detail in the published study.

In terms of model evaluation, the predicted output values for given sequences were evaluated after each training cycle. As an example, network output was considered to be predicting correctly if output values correlated to observed antigenic activity and fell within specified deviations: ±5% in anti-HCV negative samples or a maximum of ±25% in anti-HCV positive samples. Performance was based on overall predictions, obtained by averaging model prediction performance measures (specificity, sensitivity, accuracy, and correlation coefficient) from an iterative leave-one-out cross-validation (LOOCV) testing of all 12 NS3 variants (see Figure 5.6 for histograms). In LOOCV, each training example is labeled by a classifier trained on all other training examples. Here, test sets of one sample are selected, and the accuracy of the model derived from the remaining $(n - 1)$ samples is tallied. The predictive error achieved as a result is used as an appraisal of internal validation of the

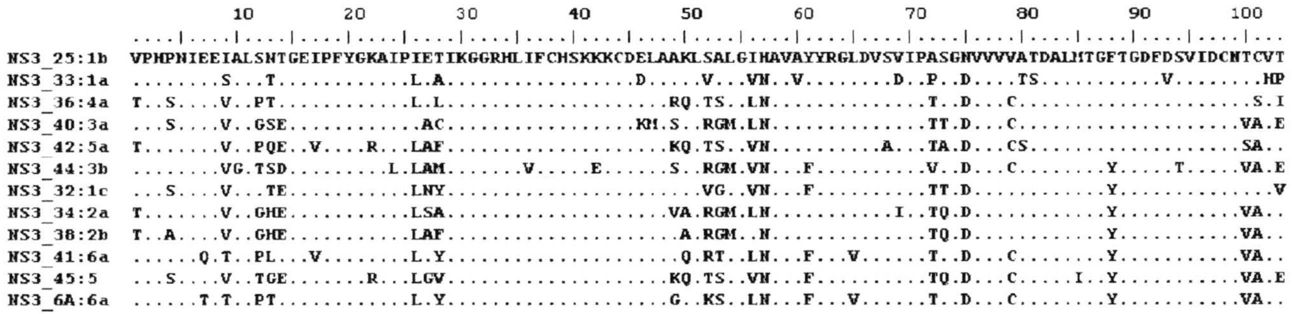

FIGURE 5.5 HCV NS3 sequences utilized in Lara et al. (2008). These sequences were initially transformed into 309-dimensional input vectors using three defined physiochemical property scales (hydrophobicity, volume, and polarity) before neural network modeling. (From Lara et al. 2008. *Bioinformatics* 24: 1858–1864. With permission from Oxford University Press.)

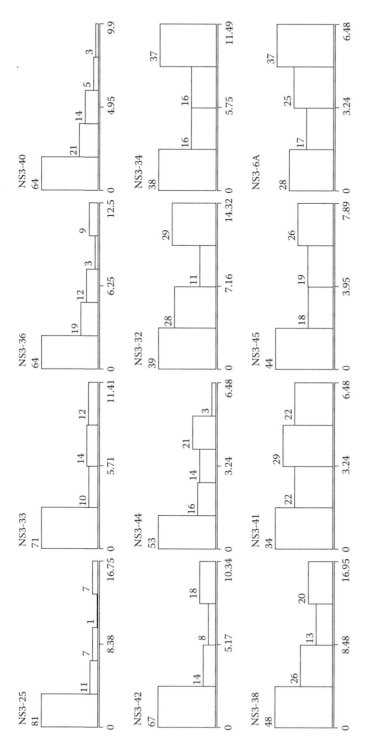

FIGURE 5.6 Associated histograms for antigenic profiles of all 12 HCV NS3 variants. Shown are the observed enzyme immunoassay (EIA) signal/cutoff (S/Co) values (x-axis), and respective number of serum samples to which variants reacted (y-axis). (From Lara et al. 2008. *Bioinformatics* 24: 1858–1864. With permission from Oxford University Press.)

TABLE 5.3

Overall Performance of a Neural Network Based on Different Feature Schemes and LOOCV Evaluation

Features	Specificity (%)	Sensitivity (%)	Accuracy (%)	Correlation Coefficient
3-properties[A]	68.7	76.2	75.5	0.7729
3-properties[B]	85.7	93.3	89.8	0.8114
2D propensities	59.2	63.0	62.9	0.5838
2D predicted	50.0	44.6	45.8	0.3754
3-PCA	70.3	76.5	74.1	0.7623
5-PCA	58.7	57.7	56.6	0.4498
3-properties[B] and 2D propensities	55.1	62.2	58.8	0.4155
3-properties[B] and 2D predicted	54.4	50.9	51.8	0.3737
3-properties[B] (r)	45.8	48.6	46.5	0.4019
2-properties[B] –h	69.9	75.8	72.8	0.6645
2-properties[B] –p	70.0	81.0	76.6	0.7697
2-properties[B] –v	73.6	85.1	81.7	0.7489

Source: From Lara et al. 2008. *Bioinformatics* 24: 1858–1864. With permission from Oxford University Press.

predictive influence of the classifier developed using the complete data set. Table 5.3 shows the average performance of the neural network models using the LOOCV approach with four different coding schemes of amino acid representations:

1. The combination of hydrophobicity, volume, and polarity physiochemical scales (*3-properties*[A] and *3-properties*[B])
2. The use of secondary structure information (*2D propensities* and *2D predicted*)
3. The use of the first three or five eigenvector components derived by PCA from a collection of 143 amino acid properties (*3-PCA* and *5-PCA*)
4. The combination of physicochemical scales with secondary structure information (*3-properties*[B] and *2D propensities,* and *3-properties*[B] and *2D predicted*).

Ultimately, the accuracy of the neural network QSAR model was compellingly dependent on the number and types of features used for sequence encoding and representation. As described in Chapter 3, feature selection involves determining a high-quality feature subset given a set of candidate features, a process applicable when the data set consists of more variables than could be efficiently included in the actual network model building phase. As guidance for this study, input features from the best representation scheme (*3-properties*[B]) were removed one at a time to examine their relative significance. And as with related studies, feature selection was based on the reaction of the cross-validated data set classification error due to the elimination of individual features. By and large, the optimized neural network performed quantitative predictions of the EIA and S/Co profiles from sequence information alone with 89.8% accuracy. Specific amino acid positions and physiochemical factors associated with the HCV NS3 antigenic properties were fully identified. The location of these positions (mapped on the NS3 3D structure) validated the major associations found by the neural network model between antigenicity and structure of the HCV NS3 proteins.

The development of suitable chemoinformatics strategies to be able to classify compounds according to the isoforms by which they are metabolized is a fundamental consideration in the drug discovery process, particularly in regard to the assessment of toxicity and drug interactions (Lynch and Price, 2007). Moreover, the early detection of ADMET (absorption, distribution, metabolism, elimination, and toxicity) properties of drugs under in vivo conditions is experimentally time consuming and expensive (de Groot et al., 2006). Considering the ADMET process further, cytochrome P450 (CYP450), a class of hemoprotein enzymes, plays a crucial role in the detoxification of xenobiotics and is believed to metabolize most foreign compounds including carcinogens and drugs (Lewis, 2001). Individual drugs can prospectively be metabolized by different CYP450 isoforms, which are classified according to the similarity of their amino acid sequences (Michielan et al., 2009). Thus, the prediction of the metabolic fate of drugs is central to helping alleviate/prevent drug–drug interactions, with such considerations as multiple binding sites, polymorphism, and enzyme induction factored into the detoxification process (Ingelman-Sundberg, 2004).

To explore this area of research, Michielan et al. (2009) developed and compared single-label and multilabel classification schemes toward the prediction of the isoform specificity of cytochrome P450 substrates (CYP450 1A2, 2C19, 2C8, 2C9, 2D6, 2E1, and 3A4). The data set consisted of 580 substrates with varying chemical structures and literature sources (see Appendix III for details). Of the 580 listed compounds, 488 substrates were metabolized by one CYP450 isoform, with the remaining being metabolized by several (up to five) isoforms. Upon extraction from the literature, the correct stereochemistry was assigned to the substrates. For modeling purposes, three different data sets were assembled and split into training, validation, and test sets. The data sets were as follows:

1. **Data Set 1**: 580 chemically distinct substrates metabolized by the seven CYP450 isoforms and used to perform multilabel classification analysis;
2. **Data Set 2**: 554 chemical structures metabolized by five different isoforms (CYP1A2, CYP2C9, CYP2D6, CYP2E1, and CYP3A4 single- and multilabel substrates). Note that all CYP2C19 and CYP2C8 single-label substrates were removed. A multilabel classification approach was used.
3. **Data Set 3**: 484 chemical structures applied to perform a single-label classification analysis.

A variety of computational methodologies were applied to build the highlighted classification models, including molecular structure building, autocorrelation function transforms, variable selection, and molecular descriptor calculations. For the latter, various combinations of molecular descriptors (Table 5.4) were selected by the

TABLE 5.4
List of Descriptors (Arranged by Class) Calculated for Single-Label and Multilabel Classification Analyses

Number	Name	Description/Details
		Global
1	MW	Molecular weight
2	HAccPot	Highest hydrogen-bond acceptor potential
3	HDonPot	Highest hydrogen-bond donor potential
4	HAcc	Number of hydrogen-bonding acceptors derived from the sum of nitrogen and oxygen atoms in the molecule
5	HDon	Number of hydrogen-bonding donors derived from the sum of NH and OH groups in the molecule
6	TPSA	Topological polar surface area
7	ASA	Approximate surface area
8	α	Mean molecular polarizability
9	μ	Molecular dipole moment
		Topological
10,11	χ^0, χ^1	Connectivity χ indices
12,13	κ_1, κ_2	κ shape indices
14	W	Wiener path number
15	χ^R	Randic index
		Size/Shape
16	D_3	Diameter
17	R_3	Radius
18	I_3	Geometric shape coefficient
19	r_2	Radius perpendicular to D_3
20	r_3	Radius perpendicular to D_3 and r_2
21–23	$\lambda_1, \lambda_2, \lambda_3$	Principal moments of inertia
24	r_{gyr}	Radius of gyration
25	r_{span}	Radius of the smallest sphere, centered at the center of mass that completely encloses all atoms in the molecule
26	ε	Molecular eccentricity
27	Ω	Molecular asphericity
		Functional-Group Counts
28	n_{aliph_amino}	Number of aliphatic amino groups
29	n_{aro_amino}	Number of aromatic amino groups
30	n_{prim_amino}	Number of primary aliphatic amino groups
31	n_{sec_amino}	Number of secondary aliphatic amino groups
32	n_{tert_amino}	Number of tertiary aliphatic amino groups
33	$n_{prim\_sec\_amino}$	$n_{prim\_amino} + n_{sec\_amino}$
34	$n_{aro\_hydroxy}$	Number of aromatic hydroxy groups
35	$n_{aliph\_hydroxy}$	Number of aliphatic hydroxy groups

(Continued)

TABLE 5.4 CONTINUED

List of Descriptors (Arranged by Class) Calculated for Single-Label and Multilabel Classification Analyses

Number	Name	Description/Details
36	$n_{guanidine}$	Number of guanidine groups
37	$n_{basic\_nitrogen}$	Number of basic, nitrogen-containing functional groups
38	$n_{acidic\text{-}groups}$	Number of acidic functional groups
39	$n_{acylsulfonamides}$	Number of sulfonamide-C=O groups
40	$n_{enolate\_groups}$	Number of enolate groups
		Vectorial
41–51	2D-ACχ^{LP}	Topological autocorrelation; property: lone-pair electronegativity χ^{LP}
52–62	2D-ACχ^{σ}	Topological autocorrelation; property: σ-electronegativity χ^{σ}
63–73	2D-ACχ^{π}	Topological autocorrelation; property: π-electronegativity χ^{π}
74–84	2D-ACq^{σ}	Topological autocorrelation; property: σ-charge q^{σ}
85–95	2D-ACq^{π}	Topological autocorrelation; property: σ-charge q^{π}
96–106	2D-ACq^{tot}	Topological autocorrelation; property: total charge q^{tot}
107–117	2D-AC$_{\alpha}$	Topological autocorrelation; property: polarizability α
118–245	3D-AC$_{identity}$	Spatial autocorrelation; property: identity
246	$\chi_{\sigma\_}1 = \Sigma\chi_{\sigma}^2$	Property: σ-electronegativity χ^{σ}
247	$\chi_{\pi\_}1 = \Sigma\chi_{\pi}^2$	Property: π-electronegativity χ^{π}
248	$q_{\sigma\_}1 = \Sigma q_{\sigma}^2$	Property: σ-charge q^{σ}
249	$q_{\pi\_}1 = \Sigma q_{\pi}^2$	Property: π-charge q^{π}
250–261	SurfACorr_ESP	Spatial autocorrelation; property: molecular electrostatic potential
262–273	SurfACorr_HBP	Spatial autocorrelation; property: hydrogen-bonding potential

Source: From Michielan et al. 2009. *Journal of Chemical Information and Modeling* 49: 2588–2605. With permission granted by the American Chemical Society.

authors to compute classification models. The size and number of molecules under examination can vary considerably; consequently, the computational requirements to perform the calculations fluctuate greatly. To a large extent, the listed descriptors in this study are 2D and 3D molecular descriptors and reflect the properties of shape and reactivity, as well as containing information pertaining to size, symmetry, and atom distribution. 2D descriptors are centered on the molecular connectivity of the compounds under study, whereas 3D descriptors are computed from a three-dimensional structure of the compounds (Kier et al., 1975; Cramer et al., 1988). The descriptors reflecting molecular shape or the distribution of a property on the molecular surface required the previous computation of 3D molecular structures. More detailed narrative of descriptor calculations and associated derivations can be found in the original study (Michielan et al., 2009).

The authors used a variety of modeling techniques in combination with the descriptors presented in Table 5.4. For multilabel classification, cross-training support vector machine (ct-SVM), multilabel k-nearest-neighbor (M_L-kNN), and counter-propagation neural network (CPG-NN) analyses were used. Simple logistic regression and SVM algorithms were applied as modeling methods for single-label classification. Especially important for the focus of this book is the CPG-NN approach, although the original study gave equal exposure and importance to the other listed models. A CPG-NN is a natural extension of Kohonen's SOM, where specified output layers (corresponding to Y variables) are added to Kohonen input layers (corresponding to molecular descriptors or X variables). CPG-NNs are qualified for developing a mathematical model capable of recognizing the class membership of each object. Moreover, there is formally no limit to the dimension of representation vectors; thus, CPG-NNs are particularly useful in QSAR studies, where the number of descriptors typically overwhelms the number of objects (Vracko, 2005). A flow diagram of the seven output layer CPG-NN used in this study is presented in Figure 5.7. As validation, LOOCV and fivefold cross-validation were employed. In the latter, the sample set was divided into fifths; one-fifth was used as a test set, and the learner was trained on the remaining. This process was repeated with a different fifth used for testing each time, with the average error rate considered. The parameter's true positive rate (TP rate), the false positive rate

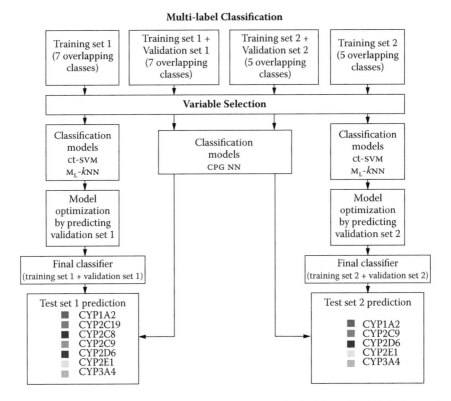

FIGURE 5.7 A flow diagram of the seven output layer CPG-NN used in Michielan et al. (2009). (From Michielan et al. 2009. *Journal of Chemical Information and Modeling* 49: 2588–2605. With permission from the American Chemical Society.)

(FP rate), the true negative rate (TN rate), the false negative rate (FN rate), the recall, and the precision were calculated for model evaluation (and comparison of multilabel and single-label approaches). As a guide, low FP and FN rates and high TP and TN values indicated good modeling performance. Additionally, Matthews correlation coefficients (MCCs) were calculated to compare the prediction results of multilabel and single-label models on given test sets. An MCC is an evaluation of the predictive performance that incorporates both precision and recall into a single, reportable value.

As standard practice in neural network model applications, the authors used a variety of different test sets for model validation. They were as follows:

1. **Test Set 1**: 217 compounds metabolized by seven different CYP450 iso-forms analyzed by the ct-SVM/7*classes*
2. **Test Set 2**: 209 substrates analyzed by both the ct-SVM/5*classes* and CPG-NN/5*classes* isoform predictors
3. **Test Set 3**: 191 CYP450 substrates analyzed by both SimLog/5*classes* and SVM/5*classes* models

A comprehensive examination of test results is outside the scope of this section (and focus of this book), but readers are encouraged to consult the original study for more details. However, clear evidence of model performance is worthy of mention. The best model (ct- SVM/5*classes*) was derived after investigator selection of 27 descriptors and reported to yield 77.5%–96.6% correct predictions for the five classes of the corresponding test set. Similarly, the CPG-NN/5*classes* model achieved 75.6%–97.1% correct predictions. In terms of single-label classification, a five-class data set was used in combination with automatic variable selection. The highest test set predictivity (78% correct) was achieved by the SVM/5*classes* model based on 19 descriptors. In the end, all models showed acceptable performances, but the multilabel predictions results were reported to more clearly reflect the real metabolic fate of the drugs studied. The authors claim that the developed multilabel methodology might well prove beneficial in exploring the metabolic profiles of new chemical entities, but stress that its prediction ability may be improved with the collection of additional multilabel substrates to add to the existing database.

5.2.3 PSYCHOLOGICAL AND PHYSICAL TREATMENT OF MALADIES

Webb et al. (2009) used a combination of a three-factor experimental design and an MLP for the optimization of gradient elution HPLC separation of nine benzodiaz-epines (nitrazepam, oxazepam, alprazolam, flunitrazepam, temazepam, diazepam, 7-aminoflunitrazepam, 7-aminonitrazepam, and 7-aminoclonazepam) in postmortem samples. Benzodiazepines are among the most commonly prescribed depressant medications today, with more than 15 different types on the market to treat a wide array of both psychological and physical maladies based on dosage and implications. In this study, 15 initial experiments based on a central composite design (CCD) were performed (Table 5.5), of which nine were assigned as training points and six as

TABLE 5.5

Experimental Values for the Three-Factor Central Composite Design Used for Data Input for ANN Optimization of HPLC Separation of Nine Benzodiazepines

Experiment	ACN grad	%MeOH	Initial %ACN
1	1	20	7.5
2	2	10	7.5
3	3	20	7.5
4	2	30	7.5
5	2	20	10
6	2	20	5
7	2	20	7.5
8	3	30	10
9	3	10	10
10	1	30	10
11	1	10	10
12	3	30	5
13	3	10	5
14	1	30	5
15	1	10	5

Source: From Webb et al. 2009. *Journal of Chromatography B* 877: 615–662. With permission from Elsevier.

verification points. CCDs contain embedded factorial or fractional factorial designs with center points that are augmented with a group of axial (star) runs that allow estimation of curvature (Hanrahan et al., 2008). More specifically, CCDs consist of cube points at the corners of a unit cube that is the product of the intervals [−1, 1], star points along the axes at or outside the cube, and center points at the origin. For this study, each experiment was performed in replicate with 10 μL injection of a mixed aqueous standard containing the nine benzodiazepines. Neural network architectures were constructed with training, and verification errors assessed.

To determine the predictive ability of the selected neural network (MLP 3:3-12-8:8), observed peak resolution values obtained from the 15 experiments were plotted against the resolution values predicted by the model, with good predictive ability shown ($r^2 = 0.9978$). A second neural network was constructed (MLP 3:3-12-8:8) using the average resolution values of each replicate. As before, good predictive ability was shown ($r^2 = 0.9968$). Using the best-performing model (trained on replicate retention time data), the authors predicted the optimum experimental conditions: 25 mM formate buffer (pH 2.8), 10% MeOH, and CAN gradient 0–15 min, 6.5%–48.5%. The optimized method was validated for real blood samples, ultimately being applied to authentic postmortem samples (Figure 5.8). The reported limits of detection ranged from 0.0057 to 0.023 μg mL^{-1}, and recoveries were on the order of 58%–92%.

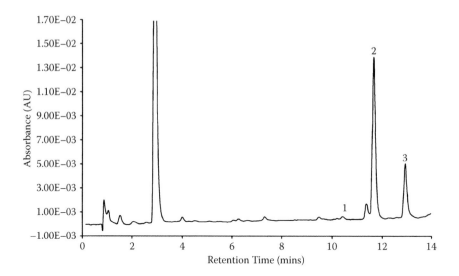

FIGURE 5.8   Representative chromatogram for postmortem blood samples. The elution order is as follows: (1) oxazepam, (2) temazepam, and (3) diazepam. (From Webb et al. 2009. *Journal of Chromatography B* 877: 615–662. With permission from Elsevier.)

Of significant interest in this study was the reported error associated with the prediction of retention time, specifically in the case of 7-aminonitrazepam. As the authors state, and as was further corroborated by Zakaria et al. (2003), predictive errors when using neural networks (and models in general) should not be less than 5% for optimization purposes. Errors associated with the prediction of the retention time for 7-aminonitrazepam were upward of 7.12%, but reported to have no bearing on the final outcome as 7-amino metabolites were well resolved under all experimental conditions. In the end, the combination of experimental designs and neural network modeling proved valuable in the optimization of a gradient elution HPLC method for nine selected benzodiazepines.

5.2.4   PREDICTION OF PEPTIDE SEPARATION

Traditionally, proteomics investigators have relied on trial-and-error methods of optimization when performing peptide separations by common chromatographic techniques. Although this approach is standard practice, it limits the prediction of the types of peptides that will be present in a given chromatographic fraction using peptide sequence information. New informatics tools that enable the investigator to predict whether a peptide will be retained on a column would thus be welcomed by the scientific community. As a prime example, Oh et al. (2007) reported the first known use of neural networks in modeling the prediction of peptide behavior on-column. More specifically, a multilayer perceptron (used as a pattern classifier) combined with a genetic algorithm (used for training the network) was used to predict whether a given peptide, based on peptide sequence, will be retained on a strong anion exchange (SAX) chromatography column. Fourteen proteins were digested and then analyzed on the SAX column followed by reversed-phase chromatography with mass spectrometry detection (RP-MS). A total of 150 peptides were identified.

The authors were deliberate in regard to feature extraction: a process of mapping the original features (measurements) into a smaller quantity of features that include the important information of the data structure. In this study, six features were extracted from the peptide sequence: molecular weight (f1), sequence index (f2), length (f3), N-pK_a value (f4), C-pK_a value (f5), and charge (f6). Each feature value was normalized using the following formula:

$$\text{Normalized} = \frac{2 \times (\text{raw} - \text{min})}{\text{max} - \text{min} - 1} \tag{5.1}$$

In this equation, "raw" and "normalized" refer to values of a given feature before and after normalization, respectively. The terms "min" and "max" refer to the minimum and maximum values of the corresponding feature category, respectively. The multilayer perceptron network used in this study is shown in Figure 5.9. Extensive training was performed using a genetic algorithm instrumental in minimizing the cost function f with form

$$\frac{1}{1+f}$$

Details of the typical genetic algorithm approach, one that was followed closely in this study, can be found in Chapter 4 of this book. To minimize the classification error, the authors chose the chromosome that gave the minimum testing error.

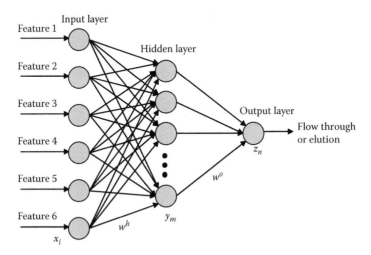

FIGURE 5.9 The multilayer perceptron used as a pattern classifier for the prediction of peptide separation in SAX chromatography. (From Oh et al. 2007. *Bioinformatics* 23: 114–118. With permission from Oxford University Press.)

Performance evaluation was carried out by first dividing the whole peptide set into two groups: a designing set (used to determine network parameters) and validation set (used to solely evaluate the performance of the network). The designing set was further divided into a training set (90 peptides) and a testing set (30 peptides). The remaining 30 were designated for the validation set. The RMSE was used to assess the training and testing errors. Additionally, classification success rates (SRs) were measured during the validation procedure. The success rate performance parameter indicates how often the neural network correctly identifies patterns of interest. The importance of each feature (and combination of features) was investigated in terms of sensitivity of the features as follows:

$$S (A|B) = SR (B) - SR (B - A) \qquad (5.2)$$

where A = the set of features that require sensitivity and B = the set of features used as inputs for classifier construction (which contains A). The investigators used the following combination as an example for demonstration purposes: the sensitivity of the combination of features 1 and 2 in the classifier constructed with features 1, 2, 3, 4, 5, 6 was

$$S (1, 2|1, 2, 3, 4, 5, 6) = SR (1, 2, 3, 4, 5, 6) - SR (3, 4, 5, 6) = 0.79 - 0.76 = 0.3$$

The authors defined the sensitivity metric as a measure of the deterioration of classification performance when chosen features are eliminated from the neural network classifier. Features with higher metric were considered more significant. Full classification results (in terms of RMSE and SR) with selected features are presented in Table 5.6. As shown in the table, the 1, 2, 3, 6 classifier displayed the best performance out of all the 15 neural network classifiers constructed using a total of four features. Additionally, they were careful to point out that in all of the classifiers that contained f4 (N-pK_a value) and f5 (C-pK_a value), the sensitivity S (4, 5|X) exhibited negative values that were therefore removed from the input layer for enhanced model performance.

Ultimately, the order of significance with respect to sensitivity was as follows: f2 (sequence index), f6 (charge), f1 (molecular weight), f3 (length), f4 (N-pK_a value), and f5 (C-pK_a value). An example method employed for analyzing the prediction performance of the neural network involved the study of the dependency of the classification success rate on molecular weight and calculated charge of peptides. The number of peptides that were correctly classified divided by the total number of peptides in the same group defined the classification success rate. As confirmed by Figures 5.10 and 5.11, no obvious correlation between the classification success rate and peptide molecular weight or peptide charge was shown. In the end, the authors reported an average classification success rate of 84% in the prediction of peptide separation on a SAX column using six features to describe individual peptides.

5.3   CONCLUDING REMARKS

This chapter presented a selection of promising studies that relate data and associated variables and describe the modeling and analysis of a variety of dynamic and

TABLE 5.6

Peptide Separation Classification Results Using Selected Features

Features	RMSE	SR
1, 2, 3, 4, 5, 6	0.76	0.79
1, 2, 3, 4, 5	0.80	0.76
1, 2, 3, 4, 6	0.70	0.82
1, 2, 3, 5, 6	0.71	0.80
1, 2, 4, 5, 6	0.74	0.78
1, 3, 4, 5, 6	0.80	0.77
2, 3, 4, 5, 6	0.78	0.78
1, 2, 3, 4	0.73	0.80
1, 2, 3, 5	0.74	0.80
1, 2, 3, 6	0.66	0.84
1, 2, 4, 5	0.79	0.78
1, 2, 4, 6	0.70	0.83
1, 2, 5, 6	0.70	0.82
1, 3, 4, 5	0.98	0.69
1, 3, 4, 6	0.76	0.80
1, 3, 5, 6	0.78	0.78
1, 4, 5, 6	0.77	0.79
2, 3, 4, 5	0.80	0.78
2, 3, 4, 6	0.74	0.81
2, 3, 5, 6	0.73	0.80
2, 4, 5, 6	0.92	0.70
3, 4, 5, 6	0.83	0.76
1, 2, 3	0.67	0.84
1, 2, 4	0.71	0.81
1, 2, 5	0.73	0.81
1, 2, 6	0.64	0.84
1, 3, 4	0.98	0.66
1, 3, 5	0.94	0.69
1, 3, 6	0.75	0.81
1, 4, 5	0.97	0.68
1, 4, 6	0.74	0.80
1, 5, 6	0.77	0.79
2, 3, 4	0.75	0.81
2, 3, 5	0.72	0.81
2, 3, 6	0.70	0.84
2, 4, 5	0.89	0.71
2, 4, 6	0.88	0.71
2, 5, 6	0.88	0.70
3, 4, 5	0.98	0.67
3, 4, 6	0.80	0.78
3, 5, 6	0.83	0.75
4, 5, 6	0.92	0.69

Source: From Oh et al. (2007). *Bioinformatics* 23: 114–118. With permission from Oxford University Press.

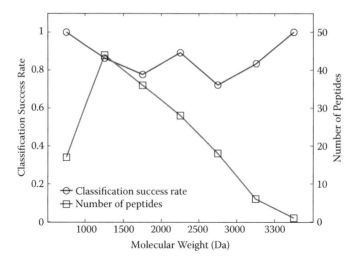

FIGURE 5.10   Plot of the correlation between the classification success rate and the molecular weight of peptides. (From Oh et al. 2007. *Bioinformatics* 23: 114–118. With permission from Oxford University Press.)

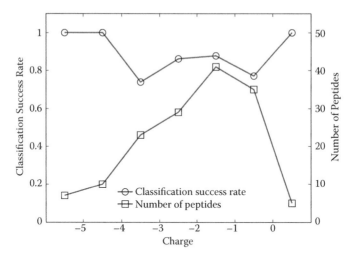

FIGURE 5.11   Plot of the correlation between the classification success rate and the charge of peptides. (From Oh et al. 2007. *Bioinformatics* 23: 114–118. With permission from Oxford University Press.)

reactive biological systems and processes involving both discrete and continuous behaviors. Many of the theoretical concepts presented in previous chapters were incorporated to guide the applications, and to foster knowledge about the continued development and refinement of neural models for areas such as bioinformatics and computational biology. We have seen the utility of neural network tools in solving complex, and often ill-posed, biological problems through data and dimensionality reduction, classification, prediction, and nonlinear modeling, with the removal of many of the previously encountered computational bottlenecks. Examples have shown the methodology of model learning and adapting in refining knowledge of model input parameters and yielded insight into the degree of obtained predictions and uncertainty for modeled output. However, it is inevitable that the issues of uncertainty regarding model accuracy and transparency will remain a common criticism among investigators and reviewers alike. But as our approach becomes more integrated (e.g., the development of hybrid models and computing technology), I believe the use of neural network models will be increasingly widespread in the biological and biomedical communities, and provide considerable insight into many real-life problems faced by modern society.

REFERENCES

Buciński, A., Socha, A., Wnuk, M., Baczek, T., Nowaczyk, A., Krysiński, J., Goryński, K., and Koba, M. 2009. Artificial neural networks in prediction of antifungal activity of a series of pyridine derivatives against *Candida albicans*. *Journal of Microbiological Methods* 76: 25–29.

Burke, B.J., and Reginer, F.E. 2003. Stopped-flow enzyme assays on a chip using a microfabricated mixer. *Analytical Chemistry* 75: 1786–1791.

Burnham, K.P., and Anderson, D.R. 2004. Multimodel inference: Understanding AIC and BIC in model selection. *Sociological Methods and Research* 33: 261–304.

Busam, S., McNabb, M., Wackwitz, A., Senevirathna, W., Beggah, S., van der Meer, J.R., Wells, M., and Harms, H. 2007. Artificial neural network study of whole-cell bacterial bioreceptor response determined using fluorescence flow cytometry. *Analytical Chemistry* 79: 9107–9114.

Carlucci, F., Tabucchi, A., Aiuti, A., Rosi, F., Floccari, F., Pagani, R., and Marinello, E. 2003. Capillary Electrophoresis in Diagnosis and Monitoring of Adenosine Deaminase Deficiency. *Clinical Chemistry* 49: 1830–1838.

Cherkasov, A., Hilpert, K., Jenssen, H., Fjell, C.D., Waldbrook, M., Mullaly, S.C., Volkmer, R., and Hancock, R.E.W. 2008. Use of artificial intelligence in the design of small peptide antibiotics effective against a broad spectrum of highly antibiotic-resistant superbugs. *ACS Chemical Biology* 4: 65–74.

Cramer, R.D., Patterson, D.E., and Bunce, J.D. 1988. Comparative molecular field analysis (CoMFA): 1. Effect of shape on binding of steroids to carrier proteins. *Journal of the American Chemical Society* 110: 5959–5967.

de Groot, M.J. 2006. Designing better drugs: Predicting Cytochrome P450 metabolism. *Drug Discovery Today* 11: 601–606.

Drug Interaction Table. Cytochrome P450. http://medicine.iupui.edu/clinpharm/ddis/table.asp (accessed February 10, 2008).

Fabry-Asztalos, L., Andonie, R., Collar, C.J., Abdul-Wahid, S. and Salim, N. 2008. A genetic algorithm optimized fuzzy neural network analysis of the affinity of inhibitors for HIV-1 protease. *Bioorganic & Medicinal Chemistry* 15: 2903–2911.

Guha, R., Stanton, D.T., and Jurs, P.C. 2005. Interpreting computational neural network quantitative structure–activity relationship models: A detailed interpretation of the weights and biases. *Journal of Chemical Information and Modeling* 45: 1109–1121.

Hanrahan, G., and Gomez, F.A. 2010. *Chemometric Methods in Capillary Electrophoresis*. John Wiley & Sons: Hoboken, NJ.

Hanrahan, G., Montes, R., and Gomez, F.A. 2008. Chemometric experimental design based optimization techniques in capillary electrophoresis: A critical review of modern applications. *Analytical and Bioanalytical Chemistry* 390: 169–179.

Havel, J., Peña-Méndez, E.M., Rojas-Hernández, A., Doucet, J.-P., and Panaye, A. 1998. Neural networks for optimization of high-performance capillary zone electrophoresis methods: A new method using a combination of experimental design and artificial neural networks. *Journal of Chromatography A* 793: 317–329.

Ingelman-Sundberg, M. 2004. Human drug metabolizing Cytochrome P450 enzymes: Properties and polymorphisms. *Biomedical and Life Sciences* 369: 89–104.

Jalali-Heravi, M., and Asadollahi-Baboli, M. 2009. Quantitative structure-activity relationship study of serotonin (5-HT7) receptor inhibitors using modified ant colony algorithm and adaptive neuro-fuzzy interference system (ANFIS). *European Journal of Medicinal Chemistry* 44: 1463–1470.

Jalali-Heravi, M., Asadollahi-Baboli, M., and Shahbazikhah, P. 2008. QSAR study of heparanase inhibitors activity using artificial neural networks and Levenberg-Marquardt algorithm. *European Journal of Medicinal Chemistry* 43: 548–556.

Kachrimanis, K., Rontogianni, M., and Malamataris, S. 2010. Simultaneous quantitative analysis of mebendazole polymorphs A-C in power mixtures by DRIFTS spectroscopy and ANN modeling. *Journal of Pharmaceutical and Biomedical Analysis* 51: 512–520.

Kaiser, D., Terfloth, L., Kopp, S., Schulz, J., de Laet, R., Chiba, P., Ecker, G.F., and Gasteiger, J. 2007. Self-organizing maps for identification of new inhibitors of P-glycoprotein. *Journal of Medical Chemistry* 50: 1698–1702.

Khudyakov, Y.E., Khudyakova, N.S., Jue, D.L., Lambert, S.B., Fang, S., and Fields, H.A. 1995. Linear B-cell epitopes of the NS3-NS4-NS5 proteins of the hepatitis C virus as modeled with synthetic peptides. *Virology* 206: 666–672.

Kier, L.B., Murray, W.J., and Hall, L.H. 1975. Molecular connectivity. 4. Relationships to biological analysis. *Journal of Medicinal Chemistry* 18: 1272–1274.

Kwak, E.S., Esquivel, S., and Gomez, F.A. 1999. Optimization of capillary electrophoresis conditions for in-capillary enzyme-catalyzed microreactions. *Analytica Chimica Acta* 397: 183–190.

Lara, J., Wohlhueter, R.M., Dimitrova, Z., and Khudyakov, Y.E. 2008. Artificial neural network for prediction of antigenic activity for a major conformational epitope in the hepatitis C virus NS3 protein. *Bioinformatics* 24: 1858–1864.

Lewis, D.F.V. 2001. *Guide to Cytochrome P450: Structure and Function*, 2nd edition. Informa Healthcare: London, U.K.

Lin, S., Arcangel, P., Medina-Selby, A., Coit, D., Ng, S., Nguyen, S., McCoin, C., Gyenes, A., Hu, G., Tandeske, L., Phelps, B., and Chien, D. 2005. Design of novel conformational and genotype-specific antigens for improving sensitivity of immunoassays for hepatitis C virus-specific antibodies. *Journal of Clinical Microbiology* 43: 3917–3924.

Livingstone, D.J. 2000. The characterization of chemical structures using molecular properties: A survey. *Journal of Chemical Information and Computational Science* 40: 195–209.

Lynch, T., and Price, A. 2007. The effect of Cytochrome P450 metabolism on drug response, interactions, and adverse effects. *American Family Physician* 76: 391–396.

Metabolite Database, MDL Inc. http://www.mdl.com/products/predictive/metabolic/index.jsp (assessed February 10, 2008).

Michielan, L. Terfloth, L., Gasteiger, J., and Moro, S. 2009. Comparison of multilabel and single-label classification applied to the prediction of the isoforms specificity of Cytochrome P450 substrates. *Journal of Chemical Information and Modeling* 49: 2588–2605.

Montes, R.E., Gomez, F.A., and Hanrahan, G. 2008. Response surface examination of the relationship between experimental conditions and product distribution in electrophoretically mediated microanalysis. *Electrophoresis* 29: 375–380.

Novotná, K., Havliš, J., and Havel, J. 2005. Optimisation of high performance liquid chromatography separation of neuroprotective peptides. Fractional experimental designs combined with artificial neural networks. *Journal of Chromatography A* 1096: 50–57.

Oh, C., Zak, S.H., Mirzaei, H., Buck, C., Regnier, F.E., and Zhang, X. 2007. Neural network prediction of peptide separation in strong anion exchange chromatography. *Bioinformatics* 23: 114–118.

Pascual-Montano, A., Donate, L.E., Valle, M., Bárcena, M., Pascual-Marqui, R.D., and Carazo, J.M. 2001. A novel neural network technique for analysis and classification of EM single-particle images. *Journal of Structural Biology* 133: 233–245.

Petritis, K., Kangas, L.J., Yan, B., Monroe, M.E., Strittmatter, E.F., Qian, W.-J., Adkins, J.N., Moore, R.J., Xu, Y., Lipton, M.S., Camp, D.G., and Smith, R.D. 2006. Improved peptide elution time prediction for reversed-phase liquid chromatography-MS by incorporating peptide sequence information. *Analytical Chemistry* 78: 5026–5039.

Rao, C.S., Sathish, T., Mahalaxmi, M., Laxmi, G.S., Rao, R.S., and Prakasham, R.S. 2008. Modelling and optimization of fermentation factors for enhancement of alkaline protease production by isolated *Bacillus circulans* using feed-forward neural network and genetic algorithm. *Journal of Applied Microbiology* 104: 889–898.

Riveros, T., Hanrahan, G., Muliadi, S., Arceo, J., and Gomez, F.A. 2009b. On-Capillary Derivatization Using a Hybrid Artificial Neural Network-Genetic Algorithm Approach. *Analyst* 134: 2067–2070.

Riveros, T., Porcasi, L., Muliadi, S., Hanrahan, G., and Gomez, F.A. 2009a. Application of artificial neural networks in the prediction of product distribution in electrophoretically mediated microanalysis. *Electrophoresis* 30: 2385–2389.

Tang, Z., Martin, M.V., and Guengerich, F.P. 2009. Elucidation of functions of human Cytochrome P450 enzymes: Identification of endogenous substrates in tissue extracts using metabolomic and isotopic labeling approaches. *Analytical Chemistry* 81: 3071–3078.

Vracko, M. 2005. Kohonen artificial neural network and counter propagation neural network in molecular structure-toxicity studies. *Current Computer-Aided Drug Design* 1: 73–78.

Webb, R., Doble, P., and Dawson, M. 2009. Optimisation of HPLC gradient separations using artificial neural networks (ANNs): Application to benzodiazepines in post-mortem samples. *Journal of Chromatography B* 877: 615–62.

Webster, G.T., de Villiers, K.A., Egan, T.J., Deed, S., Tilley, L., Tobin, M.J., Bambery, K.R., McNaughton, D., and Wood, B.R. 2009. Discriminating the intraerythrocytic lifecycle stages of the malaria parasite using synchrotron FI-IR spectroscopy and an artificial neural network. *Analytical Chemistry* 81: 2516–2524.

Wongravee, K., Lloyd, G.R., Silwood, C.J., Grootveld, M., and Brereton, R.G. 2010. Supervised self organizing maps for classification and determination of potentially discriminatory variables: Illustrated by application to nuclear magnetic resonance metabolic profiling. *Analytical Chemistry* 82: 628–638.

Wood, M.J., and Hirst, J.D. 2005. Recent Application of Neural Networks in Bioinformatics. In B. Apolloni, M. Marinaro, and R. Tagliaferri (Eds.), *Biological and Artificial Intelligence Environments*. Springer: New York.

Zakaria, P., Macka, M., and Haddad, P.R. 2003. Separation of opiate alkaloids by electrokinetic chromatography with sulfated-cyclodextrin as a pseudo-stationary phase. *Journal of Chromatography A* 985: 493–501.

Zhang, Y., Kaddis, J., Siverio, C., Zurita, C., and Gomez, F.A. 2002. On-Column Enzyme-Catalyzed Microreactions Using Capillary Electrophoresis: Quantitative Studies. *Journal of Capillary Electrophoresis and Microchip Technology* 7: 1–9.

Zhang, Y., Li, H., Hou, A., and Havel, J. 2005. Artificial neural networks based on genetic input selection for quantification in overlapped capillary electrophoresis peaks. *Talanta* 65: 118–128.

6 Applications in Environmental Analysis

6.1 INTRODUCTION

Mathematical models used to investigate the physical, chemical, and biological properties of complex environmental systems have been well received by the scientific community and are widely employed in simulation, prediction, classification, and decision-support inquiries. Of fundamental importance in modeling efforts is the capacity to categorize potentially harmful circumstances and substances and quantitatively evaluate the degree of possible consequences of catastrophic events. Models must factor in a wide range of interacting processes (both natural and anthropogenic) over varied time intervals in domains of different scales (local, mesoscale, regional, hemispherical, and global) with inherent uncertainties, sensitivity theory, and multidimensional factor analysis taken into consideration. Information acquired from comprehensive modeling efforts can be beneficial for monitoring natural environmental quality, assessing risk, developing guidelines and regulations, and planning for future economic activity. Neural network approaches in particular have shown great promise in this field, given their capacity to model complex nonlinear systems with expansive relevance to an ever-expanding array of applications. They have the ability to distinctively model contaminants both spatially and temporally across a wide assortment of matrices, act as tools for elucidating complex interrelationships and contributory mechanisms, and have displayed widespread use and applicability in environmental decision-making and management processes.

Awareness of neural modeling in the environmental field is becoming increasingly evident given the emerging number of published studies and reference works cataloged in the literature. For example, Krasnopolsky and colleagues published two significant papers underscoring the importance of neural network models in this field (Krasnopolsky and Chevallier, 2003; Krasnopolsky and Schiller, 2003). Topics addressed include critical problems in remote geophysical measurements and information on advancing computational efficiency of environmental numerical models. A special issue paper by Cherkassky et al. (2006) expanded discussion of computational intelligence in the earth and environmental sciences by introducing a generic theoretical framework for predictive learning as it relates to data-driven applications. The issues of data quality, selection of the error function, integration of predictive learning methods into the existing modeling frameworks, expert knowledge, model uncertainty, and other application-specific problems are discussed. Chau (2006) published a more targeted review highlighting the integration of artificial intelligence into coastal modeling activities. Although focused predominately on knowledge-based systems (KBSs), constructive advice on incorporating neural networks in environmental management activities is given. In a more recent document, Chen et al. (2008) discussed the suitability of neural networks and other AI-related techniques for modeling environmental systems. They provided case-specific AI applications in a wide variety of complex environmental systems. Finally, May et al. (2009) wrote an instructive review of neural networks for water quality monitoring and analysis in particular, which provides readers with guided knowledge at all stages of neural network model development and applications in which they have been found practical.

6.2   APPLICATIONS

Table 6.1 provides an overview of neural network modeling techniques in recent environmental analysis efforts. Included in the table is key information regarding application areas, model descriptions, key findings, and overall significance of the work. As with Table 5.1, this listed information is by no means comprehensive, but it does provide a representative view of the flexibility of neural network modeling and its widespread use in the environmental sciences. More detailed coverage (including model development and application considerations) of a variety of these studies, as well as others covered in the literature, is highlighted in subsequent sections of this chapter, which are dedicated to specific environmental systems or processes.

6.2.1   Aquatic Modeling and Watershed Processes

Global water and element cycles are controlled by long-term, cyclical processes. Understanding such processes is vital in the interpretation of the environmental behavior, transport and fate of chemical substances within and between environmental compartments, environmental equilibria, transformations of chemicals, and assessing the influence of, and perturbation by, anthropogenic activities. However, genuine advancement in predicting relative impacts (e.g., stability downstream and downslope from an original disruption) will require advanced integrated modeling efforts to improve our understanding of the overall dynamic interactions of these processes. For example, interactions between chemical species in the environment and aquatic organisms are complex and their elucidation requires detailed knowledge of relevant chemical, physical, and biological processes. For example, work by Nour et al. (2006) focused on the application of neural networks to flow and total phosphorus (TP) dynamics in small streams on the Boreal Plain, Canada. The continental Western Boreal Plain is reported to exhibit complex surface and groundwater hydrology due to a deep and heterogeneous glacial deposit, as well as being continually threatened by increased industrial, agricultural, and recreational development (Ferone and Devito, 2004).

Neural network model development was driven by the fact that physically based models are of limited use at the watershed scale due to the scarcity of relevant data and the heterogeneity and incomplete understanding of relevant biogeochemical processes (Nour et al., 2006). For example, the Boreal Plain houses ungauged watersheds where flow is not monitored. Development of a robust model that will effectively

TABLE 6.1
Selected Neural Network Model Applications in Modern Environmental Analysis Efforts

Analyte/Application Area	Model Description	Key Findings/Significance	Reference
Nutrient loads (N–NO$_3$ and PO$_4^{3-}$) in watersheds under time-varying human impact	Adaptive Neuro-Fuzzy Inference System (ANFIS)	The ANFIS gave unbiased estimates of nutrient loads with advantages shown over other methods (e.g., FLUX and Cohn). It allowed the implementation of a homogeneous, model-free methodology throughout the given data series	Marcé et al. (2004)
Decision support for watershed management	Neural network model trained using a hybrid of evolutionary programming (EP) and the BP algorithm	The hybrid algorithm was found to be more effective and efficient than either EP or BP alone, with a crucial role in solving the complex problems involved in watershed management	Muleta and Nicklow (2005)
Small stream flow and total phosphorus (TP) dynamics in Canada's Boreal Plain	MLP trained with a gradient descent back-propagation algorithm with batch update (BP-BU)	Four neural models were developed and tested (see Table 6.1) in Canada's Boreal Plain. Optimized models in combination with time domain analysis allowed the development of an effective stream flow model. More information about total phosphorus export is needed to fully refine the model	Nour et al. (2006)
The assessment of polychlorinated dibenzo-*p*-dioxins and dibenzofurans (PCDD/Fs) in soil, air, and herbage samples	Self-organizing map (SOM)	With the help of SOM, no significant differences in PCDD/F congener profiles in soils and herbage were noted between the baseline and the current surveys. This is an indicator that a proposed hazardous waste incinerator would not significantly impact its surrounding environment	Ferré-Huguet et al. (2006)
Modeling NO$_2$ dispersion from vehicular exhaust emissions in Delhi, India	Multilayered feedforward neural network with back-propagation	Optimized neural model used to predict 24-h average NO$_2$ concentrations at two air qualities. Meteorological and traffic characteristic inputs utilized in the model	Nagendra and Khare (2006)
Gap-filling net ecosystem CO$_2$ exchange (NEE) study	Hybrid genetic algorithm and neural networks (GNNs)	The GNN method offered excellent performance for gap-filling and high availability due to the obviated need for specialization of ecological or physiological mechanisms	Ooba et al. (2006)

(Continued)

TABLE 6.1   CONTINUED
Selected Neural Network Model Applications in Modern Environmental Analysis Efforts

Analyte/Application Area	Model Description	Key Findings/Significance	Reference
Modeling of anaerobic digestion of primary sedimentation sludge	Adaptive Neuro-Fuzzy Inference System (ANFIS)	Effluent volatile solid (VS) and methane yield were successfully predicted by ANFIS	Cakmakci (2007)
Monitoring rice nitrogen status for efficient fertilizer management	R-ANN (neural model based on reflectance selected using MLR) PC-ANN (neural model based on PC scores)	A combination of hyperspectral reflectance and neural networks was used to monitor rice nitrogen (mg nitrogen g^{-1} leaf dry weight). The performance of the PCA technique applied on hyperspectral data was particularly useful for data reduction for modeling	Yi et al. (2007)
Microbial concentrations in a riverine database	Feedforward, three-layered neural network with back-propagation	Neural network models provided efficient classification of individual observations into two defined ranges for fecal coliform concentrations with 97% accuracy	Chandramouli et al. (2007)
Forecasting particulate matter (PM) in urban areas	Hybrid combination of autoregressive integrated moving average (ARIMA) and MLP neural network	The hybrid ARIMA-neural model accurately forecasted 100% and 80% of alert and pre-emergency PM episodes, respectively	Díaz-Robles et al. (2008)
Modeling and optimization of a heterogeneous photo-Fenton process	An initial experimental design approach with resultant data fed into a multilayered feedforward neural network	A heterogeneous photo-Fenton process was optimized for efficient treatment of a wide range of organic pollutants	Kasiri et al. (2008)
Water quality modeling; chlorine residual forecasting	Input variable selection (IVS) during neural network development	Neural network models developed using the IVS algorithm were found to provide optimal prediction with significantly greater parsimony	May et al. (2008)

Exotoxicity and chemical sediment classification in Lake Turawa, Poland	SOM with unsupervised learning	SOM allowed the classification of 44 sediment quality parameters with relation to the toxicity-determining parameter (EC_{50} and mortality). A distinction between the effects of pollution on acute chronic toxicity was also established	Tsakovski et al. (2009)
Determination of endocrine disruptors in food	Fractional factorial design combined with a MLP trained by conjugate gradient descent	The use of experimental design in combination with neural networks proved valuable in the optimization of the matrix solid-phase dispersion (MSPD) sample preparation method for endocrine disruptor determination in food	Boti et al. (2009)
Carbon dioxide (CO_2) gas concentration determination using infrared gas sensors	The Bayesian strategy employed to regularize the training of the BP ANN with a Levenberg–Marquardt (LM) algorithm	The results showed that the Bayesian regulating neural network was efficient in dealing with the infrared gas sensor, which has a large nonlinear measuring range and provided precise determination of CO_2 concentrations	Lau et al. (2009)
Mercury species in floodplain soil and sediments	Self-organizing map (SOM)	SOM evaluation allowed identification of moderately (median 173-187 ng g^{-1}, range 54–375 ng g^{-1} in soil and 130 ng g^{-1} in sediment) and heavily polluted samples (662 ng g^{-1}, range 426–884 ng g^{-1})	Boszke and Astel (2009)
Electrolysis of wastes polluted with phenolic compounds	Simple and stacked feedforward networks with varying transfer functions	Chemical oxygen demand was predicted with errors around 5%. Neural models can be used in industry to determine the required treatment period, and to obtain the discharge limits in batch electrolysis	Piuleac et al. (2010)

predict flow (and TP dynamics) in such a system would thus be well received. For this study, two watersheds (1A Creek, 5.1 km^2 and Willow Creek, 15.6 km^2) were chosen, and daily flow and TP concentrations modeled using neural networks. A data preprocessing phase was first used to make certain that all data features were well understood, to identify model inputs, and to detect possible causes of any unexpected features present in the data. Five key features identified in this study included (1) an annual cyclic nature, (2) seasonal variation, (3) variables highly correlated with time, (4) differing yearly hydrographs reflecting high rain events (in contrast to those merely dictated by snowmelt and base flow condition), and (5) noticeable flow and TP concentration hysteresis. In regards to flow, model inputs were divided into cause/effect inputs (rainfall and snowmelt), time-lagged inputs, and inputs reflecting annual and seasonal cyclic characteristics. In terms of TP modeling, cause/effect inputs were limited to flow and average air temperature. Table 6.2 summarizes the set of inputs used in the final model application.

Figure 6.1 displays a schematic of the optimum network architecture for all four modes investigated. Shown is the training process demonstrating how the input information propagated and how the error back-propagation algorithm was utilized within the neural architecture developed. More distinctly, two training algorithms were tested: (1) a gradient descent back-propagation algorithm that incorporated user-specified learning rate and momentum coefficients and (2) a BP algorithm with a batch update technique (BP-BM). In the batch update process, each pattern is fed into the network once, and the error is calculated for that specific pattern. The next

---

TABLE 6.2
Summary Table for All Model Inputs Used in Nour et al. (2006)

Final Model	Inputs
Model 1 (Q for Willow)	R_t, R_{t-1}, R_{t-2}, R_{t-3}, $sin(2\pi v t)$, $cos(2\pi v t)$, T_{max}, T_{mean}, T_{min}, dd_t, dd_{t-1}, dd_{t-2}, S_t, S_{t-1}, S_{t-2}
Model 2 (TP for Willow)	TP_{t-1}, $sin(2\pi v t)$, $cos(2\pi v t)$, T_{mean}, ΔQ_t,_ΔQ_{t-1},_ΔQ_{t-3}
Model 3 (Q for 1A)	R_t, R_{t-1}, R_{t-2}, $sin(2\pi v t)$, $cos(2\pi v t)$, T_{max}, T_{min}, dd_t, dd_{t-1}, S_t, S_{t-1}
Model 4 (TP for 1A)	TP_{t-1}, $sin(2\pi v t)$, $cos(2\pi v t)$, T_{mean}, ,ΔQ_t,_ ΔQ_{t-2},_ ΔQ_{t-3},_ ΔQ_{t-4}

Source: Nour et al. 2006. *Ecological Modelling* 191: 19–32. Modified with permission from Elsevier.

Note : R_t, R_{t-1}, R_{t-2}, and R_{t-3} are the rainfall in mm at lags 0 through 3; T_{max}, T_{mean}, and T_{min} represents maximum, daily mean, and minimum air temperatures in °C, respectively; dd_t, dd_{t-1}, and dd_{t-2} are the cumulative degree days at lags zero to two; S_t, S_{t-1}, and S_{t-2} are the cumulative snowfall in mm for lags 0 through two; $\Delta Q_t = (Q_t - Q_{t-1})$, ΔQ_{t-1},_ ΔQ_{t-2}, ΔQ_{t-3}, and_ΔQ_{t-4} are the daily change in flow at lags 1, 2, 3, and 4, respectively.

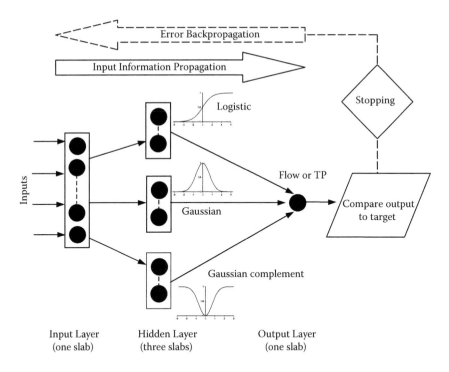

FIGURE 6.1 Schematic showing neural network optimum architecture for all four models employed in Nour et al. (2006). (From Nour et al. 2006. *Ecological Modelling* 191: 19–32. With permission from Elsevier.)

pattern is then fed, and the resultant error added to the error of the previous pattern to form a global error. The global error is then compared with the maximum permissible error; if the maximum permissible error is greater than the global error, then the foregoing procedure is repeated for all patterns (Sarangi et al., 2009). Table 6.3 presents a summary of optimum neural network model architectures and internal parameters for all four models. Model evaluation was based on four specified criteria: (1) the coefficient of determination (R^2), (2) examination (in terms of maximum root-mean-square error [RMSE]) of both measured and predicted flow hydrographs, (3) residual analysis, and (4) model stability.

Concentrating on the Willow Creek watershed, the developed flow models were shown to successfully simulate average daily flow with R^2 values exceeding 0.80 for all modeled data sets. Neural networks also proved useful in modeling TP concentration, with R^2 values ranging from 0.78 to 0.96 for all modeled data sets. Note that the R^2 value is a widely used goodness-of-fit-measure whose worth and restrictions are broadly applied to linear models. Application to nonlinear models generally leads to a measure that can lie distant from the [0,1] interval and diminish as regressors are included. In this study, a three-slab hidden layer MLP network was chosen and used for modeling with measured versus predicted flow hydrographs and the TP concentration profile presented in Figures 6.2a and 6.2b, respectively. The authors also indicated that more research on phosphorus dynamics in wetlands is necessary to

TABLE 6.3

Summary Table Showing Optimum Neural Network Model Architectures and Internal Parameters for Nour et al. (2006)

	Model 1 (Q for Willow)	Model 2 (TP for Willow)	Model 3 (Q for 1A)	Model 4 (TP for 1A)
Scaling function	Linear, $\langle\langle$-1, 1$\rangle\rangle$	Linear, $\langle\langle$-1, 1$\rangle\rangle$	Linear, $\langle\langle$-1, 1$\rangle\rangle$	Linear, $\langle\langle$-1, 1$\rangle\rangle$
Optimum network (I-HG-HL-HGC-O)	15-4-4-4-1	8-5-5-5-1	11-5-2-5-1	7-7-5-7-1
Output activation function	tanh	Logistic	tanh	tanh
Training algorithm	BP	BP-BM	BP	BP-BM
Learning rate	0.2	Insensitive	0.15	Insensitive
Momentum coefficient	0.2	Insensitive	0.15	Insensitive

Source: Nour et al. 2006. *Ecological Modelling* 191: 19–32. Modified with permission from Elsevier.

Note: I denotes the input layer; HG, HL, and HGC are the Gaussian, logistic, and Gaussian complement slabs hidden layer, respectively; tanh is the hyperbolic tangent function; and << >> denotes an open interval.

better characterize the impact of wetland areas and composition of the water phase phosphorus in neural network modeling. This is not surprising given the fact that phosphorus occurs in aquatic systems in both particulate and dissolved forms and can be operationally defined, not just as TP but also as total reactive phosphorus (TRP), filterable reactive phosphorus (FRP), total filterable phosphorus (TFP), and particulate phosphorus (PP) (Hanrahan et al., 2001). Traditionally, TP has been used in most model calculations, mainly because of the logistical problems associated with measuring, for example, FRP, caused by its rapid exchange with particulate matter.

Chandramouli et al. (2007) successfully applied neural network models to the intricate problem of predicting peak pathogen loadings in surface waters. This has positive implications given the recurrent outbreaks of waterborne and water contact diseases worldwide as a result of bacterial concentrations. Measuring the existence of pathogens in drinking water supplies, for example, can prove useful for estimating disease incidence rates. Accurate estimates of disease rates require understanding of the frequency distribution of levels of contamination and the association between drinking water levels and symptomatic disease rates. In the foregoing study, a 1,164 sample data set from the Kentucky River basin was used for modeling 44 separate input parameters per individual observation for the assessment of fecal coliform (FC) and/or atypical colonies (AC) concentrations. The overall database contained observations for six commonly measured bacteria, 7 commonly measured physico-chemical water quality parameters, rainfall and river flow measurements, and 23 input fields created by lagging flow and rainfall by 1, 2, and 3 days. As discussed in Section 3.4, input variable selection is crucial to the performance of neural network classification models. The authors of this study adopted the approach of Kim et al. (2001), who proposed the relative strength effect (RSE) as a means of differentiating

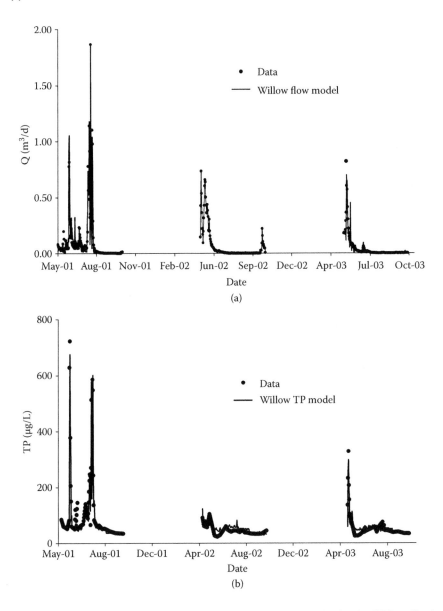

FIGURE 6.2 (a) Measured versus model predicted flow hydrographs for the Willow Creek watershed. (b) Measured versus predicted TP concentration profile from the Willow Creek watershed. (From Nour et al. 2006. *Ecological Modelling* 191: 19–32. With permission from Elsevier.)

between the relative influence of different input variables. Here, the RSE was defined as the partial derivative of the output variable y_k, $\partial y_k / \partial x_i$. If $\partial y_k / \partial x_i$ is positive, the increase in input results in an increase in output. They used the average RSE value of inputs for the p data set as training in their basic screening approach. The larger the absolute value displayed, the greater the contribution of the input variable. From the original 44 inputs, a final model (after input elimination) with 7 inputs (7:9:1 architecture) was used for predicting FC. A similar approach was applied to develop the final AC neural model (10:5:1). Table 6.4 provides the final input parameters used to model bacterial concentrations.

As discussed in Chapter 3, data sets with skewed or missing observations can affect the estimate of precision in chosen models. In this study, the authors chose to compare conventional imputation and multiple linear regression (MLR) approaches with the developed neural models. MLR is likely the simplest computational multi-variate calibration model and is typically applied when an explicit causality between dependent and independent variables is known. MLR does suffer from a number of limitations, including overfitting of data, its dimensionality, poor predictions, and the inability to work on ill-conditioned data (Walmsley, 1997). The neural network modeling approach provided slightly superior predictions of actual microbial concentrations when compared to the conventional methods. More specifically, the optimized model showed exceptional classification of 300 randomly selected, individual data observations into two distinct ranges for fecal coliform concentrations with 97% overall accuracy. This level of accuracy was achieved even without removing potential outliers from the original database. In summary, the application of the relative strength effect proved valuable in the development of precise neural network models for predicting microbial loadings, and ultimately provided guidance for the development of appropriate risk classifications in riverine systems. If the developed neural network models were coupled with a land transformation model, spatially explicit risk assessments would then be possible.

6.2.2 ENDOCRINE DISRUPTORS

It has been hypothesized that endocrine-active chemicals may be responsible for the increased frequency of breast cancer and disorders of the male reproductive tract. Synthetic chemicals with estrogenic activity (xenoestrogen) and the organochlorine environmental contaminants polychlorinated biphenyls (PCBs) and DDE have been the prime etiologic suspects (Safe, 2004). In addition, hormones naturally secreted by humans and animals have been shown to induce changes in endocrine function. Given the sizeable and expanding number of chemicals that pose a risk in this regard, there is an urgent need for rapid and reliable analytical tools to distinguish potential endocrine-active agents. A study by Boti et al. (2009) presented an experimentally designed ($3^{(4-1)}$ fractional factorial design) neural network approach to the optimization of matrix solid-phase dispersion (MSPD) for the simultaneous HPLC/UV-DAD determination of two potential endocrine disruptors: linuron and diuron (Figure 6.3) and their metabolites—1-(3,4-dichlorophenyl)-3-methylurea (DCPMU), 1-(3,4-di-chlorophenyl) urea (DCU), and 3.4-dichloroaniline (3.4-DCA)—in food samples. MSPD is a patented process for the simultaneous disruption and extraction of solid or semisolid samples, with analyte recoveries and matrix cleanup performance typically dependent on column packing and elution procedure (Barker, 2000). This procedure uses bonded-phase solid supports as an abrasive to encourage disturbance of sample architecture and a bound solvent to assist in complete sample disruption during the blending process. The sample disperses over the exterior of the bonded

TABLE 6.4

Final Selected Input Variables Used to Model Bacterial Concentrations

	Flow Lock 10	Flow Middle Fork KY[a]	Flow Red River[a]	Flow Lock 14[a]	TC[b]	BG[c]	FS[d]	FC[e]	Turbidity	Calcium hardness
FC	x	x		x	x	x	x	x	x	x
AC	x	x	x	x	x	x	x	x	x	x

Source: Chandramouli et al. 2007. *Water Research* 41: 217–227. With permission from Elsevier.

[a] One-day lagged flow value.
[b] TC = Total coliform group colonies.
[c] BG = Background colonies.
[d] FS = Fecal streptococci.
[e] FC = Fecal coliforms.

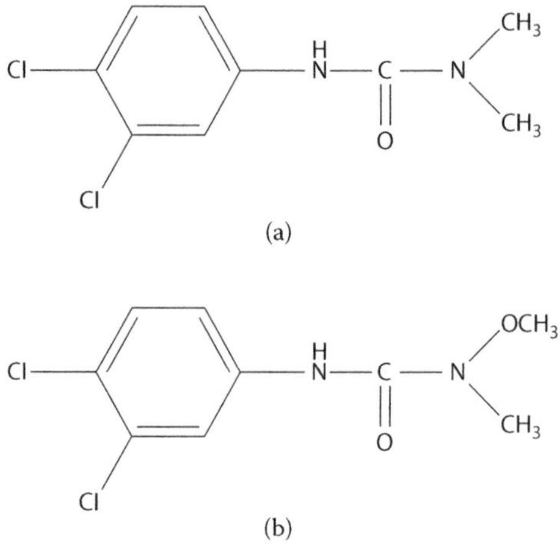

(a)

(b)

FIGURE 6.3 Two endocrine disruptors: (a) linuron and (b) diuron.

phase-support material to provide a new mixed phase for separating analytes from an assortment of sample matrices (Barker, 1998).

When combined with experimental design techniques, neural network models have been shown to provide a reduction in the number of required experiments and analysis time, as well as enhancing separation without prior structural knowledge of the physical or chemical properties of the analytes. In fractional factorial designs, the number of experiments is reduced by a number p according to a 2^{k-p} design. In the most commonly employed fractional factorial design, the half-fraction design ($p = 1$), exactly one half of the experiments of a full design are performed. It is based on an algebraic method of calculating the contributions of the numerous factors to the total variance, with less than a full factorial number of experiments (Hanrahan, 2009). For this study, the influence of the main factors on the extraction process yield was examined. The selected factors and levels chosen for the $3^{(4-1)}$ fractional factorial design used this study are shown in Table 6.5. These data were used as neural network input. Also included are the measured responses, average recovery (%), and standard deviation of recovery values (%), which were used as model outputs for neural model 1 and neural model 2, respectively. Concentrating on model 2 in detail, the final architecture was 4:10:1, with the training and validation errors at 1.327 RMS and 1.920 RMS, respectively. This resulted in a reported $r = 0.9930$, thus exhibiting a strong linear relationship between the predicted and observed standard deviation of the average recovery (%). Response graphs were generated, with the maximum efficiency achieved at 100% Florisil, a sample/dispersion material ratio of 1:1, 100% methanol as the elution system, and an elution volume of 5 mL. The final elution volume was adjusted to 10 mL to account for practical experimental observations involving clean extracts, interfering peaks, as well as mixing and column preparation functionality.

The analytical performance of the optimized MSPD method was evaluated using standard mixtures of the analytes, with representative analytical figures of merit presented in Table 6.6. Included in the table are recoveries of the optimized MSPD

TABLE 6.5

$3^{(4-1)}$ Fractional Factorial Design Data Set Used for Modeling Using Neural Networks

Run	Input 1 [Florisil[a] (%)]	Input 2 [Sample/ Dispersion Ratio]	Input 3 [Methanol[b](%)]	Input 4 [Eluent volume (mL)]	Output 1 [Average recovery[c](%)]	Output 2 [Standard deviation of recovery values[c] (%)]
1	100	1:1	100	10	71	15
2	0	1:2	50	15	104	22
3	0	1:4	50	10	107	42
4	100	1:4	0	10	67	31
5	50	1:1	100	5	103	9
6	50	1:4	50	15	64	17
7	50	1:1	0	15	82	46
8	100	1:4	50	5	95	25
9	0	1:1	100	15	35	15
10	0	1:4	100	5	117	21
11	0	1:2	100	10	87	37
12	0	1:2	0	5	114	47
13	50	1:2	100	15	82	46
14	100	1:2	100	5	84	34
15	50	1:2	0	10	109	27
16	50	1:2	50	5	101	28
17	100	1:2	0	15	65	43
18	100	1:4	100	15	102	41
19	0	1:2	0	15	105	27
20	0	1:1	0	10	85	20

(Continued)

TABLE 6.5   CONTINUED

$3^{(4-1)}$ Fractional Factorial Design Data Set Used for Modeling Using Neural Networks

Run	Input 1 [Florisil[a] (%)]	Input 2 [Sample/ Dispersion Ratio]	Input 3 [Methanol[b](%)]	Input 4 [Eluent volume (mL)]	Output 1 [Average recovery[c](%)]	Output 2 [Standard deviation of recovery values[c] (%)]
21	50	1:4	0	5	71	55
22	100	1:1	50	15	75	42
23	50	1:4	100	10	92	46
24	100	1:2	50	10	94	33
25	50	1:1	50	10	100	26
26	100	1:1	0	5	52	27
27	0	1:1	50	5	93	45

Source: Modified from Boti et al. 2009. Journal of Chromatography A. 1216: 1296–1304. With permission from Elsevier.

[a] Percentage content of Florisil in C_{18} material;

[b] Percentage content of methanol in dichloromethane;

[c] ANN_1: four inputs and one output (average recovery), ANN_2: four inputs and one output (standard deviation of recovery values).

TABLE 6.6

Analytical Figures of Merit of the MSPD Method

Compound	Recovery (RSD%)	Coefficient of Determination	Repeatability[a]	Reproducibility[a] (RSD%)	LOQ (ng g^{-1})[b]	LOD (ng g^{-1})[b]
DCPU	74(4)	0.9981	12	13	5.0	15.2
DCPMU	93(3)	0.9973	8	7	1.9	5.7
Diuron	91(2)	0.9988	7	6	2.0	6.5
Linuron	96(2)	0.9997	4	5	1.8	5.3
3.4-DCA	55(6)	0.9945	10	14	3.0	9.1

Source: Modified from Boti et al. 2009. *Journal of Chromatography A.* 1216: 1296–1304. With permission from Elsevier.

[a] $n = 6$ injections;

[b] LOD = limit of detection (S/N = 3); LOQ = limit of quantification (S/N =10); detection at 252 nm for diuron and DCPMU, 250 nm for linuron and DCPU and 248 nm for 3.4-DCA.

method using three independent 0.5 g portions of potato sample spiked at a concentration level of 100 ng g^{-1} for all compounds studied. Recoveries range from 55% to 96%, with quantification limits between 5.3 and 15.2 ng g^{-1} being achieved. Finally, a variety of spiked food commodities were tested with good MSPD extraction recoveries obtained under optimized conditions.

6.2.3 ECOTOXICITY AND SEDIMENT QUALITY

A model study by Tsakovski et al. (2009) highlighted the use of SOMs in classifying and interpreting sediment monitoring data, particularly from the standpoint of studying the relationship between ecotoxicity parameters (acute and chronic toxicity) and chemical components. Although standard methods of sediment toxicity testing are fairly well accepted (Ho et al., 2000), there is a defined need to incorporate more quantitative methods for data set treatment and modeling. SOM data interpretation is one such tool, and as discussed in Chapter 2, Section 2.3.2, this technique provides for a competitive learning approach based on unsupervised learning processes. The study site of interest included major sampling regions of Turawa Lake in the southwestern portion of Poland, with the main contamination sources being industry, agriculture, and urban and domestic sewage. Bottom sediments, accumulated in Turawa Lake for over 60 years, are reported to be mainly sapropelic muds containing up to 25% of organic matter (Wierzba and Nabrdalik, 2007). Bottom sediment data sets were obtained during a comprehensive 59 site sampling campaign in 2004. Forty-four variables were considered, including ecotoxicity parameters, pesticides, congeners, PAHs, heavy metals, as well as two physical markers of the collected sediment samples (ignition loss and fraction size). Chemical analysis of the samples included various instrumental methods compiled with different chemical and ecotoxicity variables. Both acute and chronic toxicity were determined using ToxAlert 100 and Microtox model 500 instruments and the bioluminescent bacteria *Vibrio fisheri*. Chronic toxicity was tested in the presence of the crustacean *Heterocypris incongruens*.

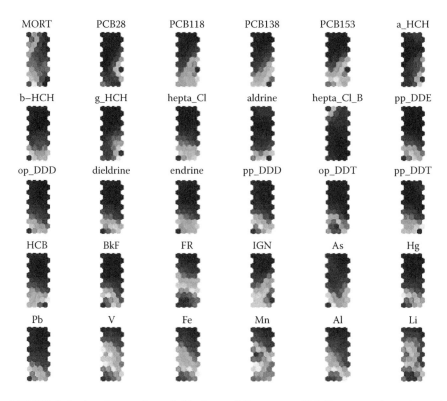

FIGURE 6.4 *A color version of this figure follows page 106.* Representative subset of SOMs for 30 of the 44 parameters tested in chronic toxicity mode. Make special note of the values from the mortality SOM, where a limited number of sites revealed high mortality. (From Tsakovski et al. 2009. *Analytica Chimica Acta* 631: 142–152. With permission from Elsevier.)

In the initial experimental setup, the authors constructed a SOM for each of the 44 sediment quality parameters (59 × 44 projection elements). Figure 6.4 provides a representative subset of SOMs for 30 of the 44 parameters tested in chronic toxicity mode. Of particular mention were the values included in the mortality SOM, which form a specific pattern in which a limited number of sites revealed a high rate of mortality. In total, four groups of sites with high levels of mortality were observed, with their vicinities presented in Figure 6.5. Group 1, consisting of nine sites, had the highest mortality index (268%). The authors assert that this was due to the high accumulation of heptachlor B, a pesticide found at the mouth of the Mala Panew River on the southeastern portion of the lake. It is one of the earliest and most widely used organochlorine pesticides whose metabolites bioaccumulate in the environment. Studies have shown that heptachlor is a liver tumor-promoter in rats and mice and encourages tumor promoting-like alterations in human myeloblastic leukemia cells (Smith, 1991). Group 2 exhibited a mortality index of nearly 200%, the contribution from heptachlor B being negligible. However, higher individual indices were exhibited for such parameters as ppDDE, dieldrine, endrine, ppDDT, PAHs, and

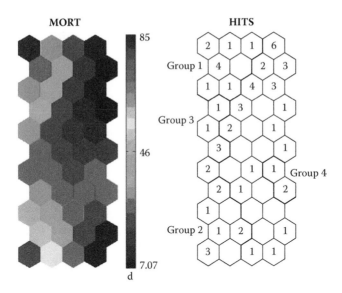

FIGURE 6.5 *A color version of this figure follows page 106.* The vicinities of four groups of sites that revealed high levels of mortality. (From Tsakovski et al. 2009. *Analytica Chimica Acta* 631: 142–152. With permission from Elsevier.)

numerous metals. Group 3, which exhibited a mortality index of nearly 140%, was characterized by relatively high indices of metals: V, Fe, Mn, Al, and Li, and a high fractional parameter. Finally, Group 4, with a mortality index of 130%, exhibited high concentrations (indices) of PCBs.

Equivalent techniques were applied to the appraisal of acute toxicity, with SOM representations of the same 30 chosen parameters highlighted in Figure 6.6. The major indicator observed was the half-maximal effective concentration (EC_{50}), whose values are indicative of the acute toxicity of the samples collected. More specifically, it refers to the concentration of a toxicant that induces a response intermediate between the baseline and maximum after a particular duration of exposure. In addition, three distinct groups of similar acute toxicity and object distribution were formed (Figure 6.7). Group 1, with an EC_{50} index of 55.2%, reflected the high danger of acute toxicity associated with heptachlor B. Group 2 exhibited a slightly higher EC_{50} (60%), which indicated a lower level of toxicity. The most significant indices exhibited were pesticides (opDDE, ppDDE, dieldrine, endrine, ppDDT, and opDDT) as well as PAHs and metals (Hg, Cd, Pb, Zn, and Cu). Group 3, with an acute toxicity index of 82.9%, featured high indices of opDDT, fraction size, V, Fe, Mn, and Li.

In both the acute and chronic toxicity studies, hit diagrams generated from the SOMs allowed the detection of content on each node for a well-determined composition. Generated SOMs made it possible to cluster the objects of interest (variables) and demonstrate their spatial proximity. Moreover, SOM data interpretation allowed an improved understanding of the discriminating tracers for each of the identified groups of similarity between sampling locations. The authors presented convincing evidence as to the usefulness of SOMs in defining a distinction between the effects

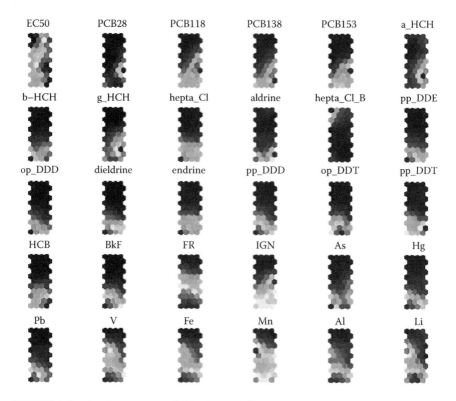

FIGURE 6.6 *A color version of this figure follows page 106.* Representative subset of SOMs for 30 of the 44 parameters tested in acute toxicity mode. The major indicator observed was EC_{50}, whose values were indicative of the acute toxicity of the collected samples. (From Tsakovski et al. 2009. *Analytica Chimica Acta* 631: 142–152. With permission from Elsevier.)

of environmental contaminants on acute and chronic toxicity based on quality sediment data.

6.2.4 Modeling Pollution Emission Processes

Primary pollutants emitted directly or indirectly into the atmosphere from a variety of sources, including nitrogen oxides (NO_x), sulfur dioxide (SO_2), carbon monoxide (CO), particulate matter (PM), and volatile organic compounds (VOCs), can be main triggers of poor air quality. A variety of human activities (e.g., combustion of fossil fuels) have also led to the emission of carbon dioxide (CO_2), a potential greenhouse gas implicated in the process of climate change. Besides the combustion of fuels, landfill gas has also been proved to be a principal source of CO_2. It is produced by the breakdown of biodegradable organic materials and comprises 30%–60% methane, 20%–50% CO_2, and other trace gases depending on the types of waste degrading (Squire and Ramsey, 2001). In response to this concern, continuous and

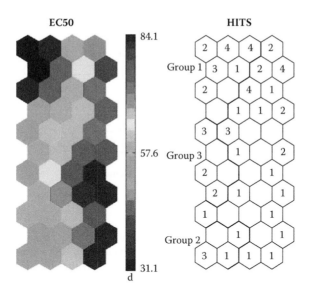

FIGURE 6.7 *A color version of this figure follows page 106.* Highlight of three distinct groups of similar acute toxicity and object distribution. (From Tsakovski et al. 2009. *Analytica Chimica Acta* 631: 142–152. With permission from Elsevier.)

automatic monitoring systems are now in place at landfills under regulation as shown by a recent study by Lau et al. (2009).

This study proved the effectiveness of overcoming the problem of overfitting during neural network training for the calibration of CO_2 gas sensors. In more specific terms, the authors employed a neural network using a Levenberg–Marquardt (LM) algorithm with Bayesian regularization to predict unknown carbon dioxide CO_2 gas concentrations in landfills via infrared (IR) gas sensor analysis. Recall from Chapter 3, Section 3.6.4, that Bayesian regularization makes use of the posterior probability of the weights and biases to obtain the best-performing neural network where overfitting is under strict control. In general, networks exhibit better generalization performance and lower susceptibility to overfitting as the network size increases. A schematic of the experimental gas sensor setup used in this study is presented in Figure 6.8. A range of CO_2 samples were obtained in-line by mixing the stock with pure dry nitrogen gas controlled by mass flow controllers as pictured. The gas mixture was then passed through a dedicated sensor chamber, collected at the outlet, and subsequently analyzed by FTIR to certify the concentration of CO_2 generated by the gas sensor. Ultimately, a measurable CO_2 gas concentration range of 0–15,000 ppm was realized.

Representative calibration curves obtained from sensor exposure to predefined CO_2 samples with increasing concentration, as well as those obtained from the same samples measured by the reference FTIR method, are presented in Figure 6.9. Also visible in the figure is the sensor calibration plot, which is a curved line with an inflection point at approximately 7000 ppm. Of obvious importance is how the sensor provided a nonlinear plot on which the dotted line is added. Given this nonlinear

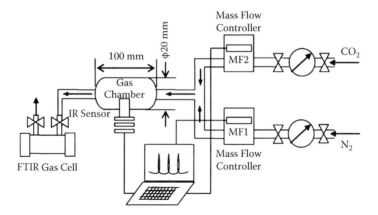

FIGURE 6.8 A detailed schematic of the experimental gas sensor setup used in predicting carbon dioxide (CO_2) gas concentrations in landfills via infrared (IR) analysis. (From Lau et al. 2009. *Sensors and Actuators B: Chemical* 136: 242–247. With permission from Elsevier.)

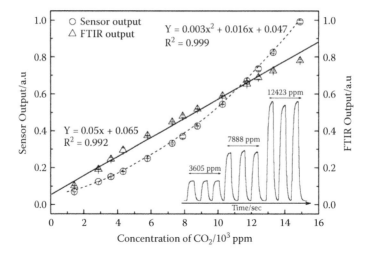

FIGURE 6.9 Representative calibration curves obtained from sensor exposure to pre-defined CO_2 samples with increasing concentration. Also shown are those obtained from the same samples measured by the reference FTIR method. (From Lau et al. 2009. *Sensors and Actuators B: Chemical* 136: 242–247. With permission from Elsevier.)

characteristic of the IR sensor to the change of concentration of CO_2, it was expected that the chosen neural approach with Bayesian regularization would prove useful in the prediction of CO_2 gas concentration. To test this assumption, the authors investigated the performance of the neural network LM algorithm with and without Bayesian regularization. Figure 6.10 presents a plot of this comparison with two hidden neurons. As shown, networks that used Bayesian regularization resulted in shorter training and testing times. Notice that the optimized neural network with LM algorithm alone resulted in better performance in regard to testing error. However,

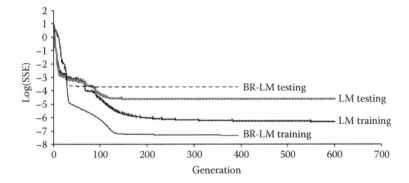

FIGURE 6.10   A plot examining the comparison of the neural network LM algorithm with and without Bayesian regularization with two hidden neurons. Networks with Bayesian regularization resulted in shorter training and testing times. (From Lau et al. 2009. *Sensors and Actuators B: Chemical* 136: 242–247. With permission from Elsevier.)

the network model incorporating Bayesian regularization was preferred, which gave a relationship coefficient of 0.9996 between targets and outputs with prediction recoveries between 99.9% and 100.0%.

Cement production is one of the most energy-intensive industrial processes and a well-documented source of contaminants, principally NO_x, sulfur oxides (SO_x), CO, CO_2, and dust, in the atmosphere (Canpolat et al., 2002; Sheinbaum and Ozawa, 1998). For example, in cement manufacturing, SO_x emission is mainly derived from the combustion of sulfur-bearing compounds in the fuels (e.g., from pyrite in coal) but can, to a less significant extent, also come from nonfuel raw materials. In order to study such releases in detail, Marengo et al. (2006) developed statistical and neural network models to predict gas emissions from a Portland cement production process. Independent model variables of interest included the chemical properties of the raw materials, the chemical and physicochemical properties of the fuels, and kiln operating conditions. The data set consisted of 365 samples taken at regular intervals over 25 successive days of January 2001 in the cement production plant of Buzzi Unicem (Italy).

In terms of statistical approaches, PCR and PLS are two multivariate regression methods used to understand and characterize chemical and physical measurements of complex systems. In PCR, the original data matrix **X** is approximated by a small set of orthogonal principal components (PCs), with the decision on which PCs to retain based on the percentages of variance that each PC explains (Hanrahan, 2009). PLS is a quintessential soft-model-based regression technique that utilizes regressing eigenvectors from the original data matrix onto the dependent variables. It is an extension of principal component analysis and employs an iterative algorithm that extracts linear combinations of essential features of the original **X** data while modeling the **Y** data dependence on the data set (Hanrahan, 2009). Concerning neural network development, two types were applied to this study: supervised (back-propagation) and unsupervised (Kohonen) networks. Recall that unsupervised networks can group objects of a data set into different classes on the basis of their similarity. Realizing this feature, the authors made use of the Kohonen approach to select the training set with the aim of finding samples that guaranteed a homogenous representation of the complete experimental domain. All 365 samples were used as input for an $8 \times 8 \times 10$ Kohonen network, with convergence achieved at 500 epochs. Samples were assigned to the cells of the top-map on the basis of the similarity of the 19 descriptors. This led to a partition of the data set into training (237 samples) and test (119 samples) sets. The training set was used for training the neural network and, in particular, for optimizing its weights by the back-propagation algorithm. The test set was used to determine the epoch to interrupt the network training. Model evaluation was performed by examination of the coefficient for multiple determination (R^2), the root-mean-square error for fitting (RMSEF), and the root-mean-square error for prediction (RMSEP).

A summary of modeled results for SO_2, dust, and NO_2 emissions are highlighted in Tables 6.7, 6.8, and 6.9, respectively. For SO_2 emissions, a performed Wilcoxon test (a nonparametric alternative to t-test for dependent samples) showed that the neural network model had the best predictive ability, while the other two methods were considered statistically equivalent. The values of RMSEP reported reflect this result. The R^2 and the RMSE for both the training and prediction for dust emissions showed similar results to those of SO_2 emissions. The neural network proved superior, with

TABLE 6.7

Summary of SO_2 Emission Results Obtained from the Three Calibration Models

Model	R^2 Training set	R^2 Prediction set	RMSEF (mg/Nm³)	RMSEP (mg/Nm³)
ANN	0.96	0.82	1.7	3.0
PCR (7 PCs)	0.69	0.51	4.1	5.1
PLS (6 LVs)	0.74	0.53	3.7	5.0

Source: Adapted from Marengo et al. 2006. *Environmental Science and Technology* 40: 272–280. With permission from the American Chemical Society.

TABLE 6.8

Summary of Dust Emission Results Obtained from the Three Calibration Models

Model	R^2 training set	R^2 prediction set	RMSEF (mg/Nm³)	RMSEP (mg/Nm³)
ANN	0.80	0.65	0.9	1.2
PCR (9 PCs)	0.58	0.35	1.4	1.7
PLS (4 LVs)	0.60	0.41	1.3	1.6

Source: Adapted from Marengo et al. 2006. *Environmental Science and Technology* 40: 272–280. With permission from the American Chemical Society.

TABLE 6.9

Summary of NO₂ Emission Results Obtained from the Three Calibration Models

Model	R^2 Training set	R^2 Prediction set	RMSEF (mg/Nm³)	RMSEP (mg/Nm³)
ANN	0.60	0.22	88.8	124.5
PCR (9 PCs)	0.28	−0.07	119.6	146.3
PLS (4 LVs)	0.36	0.01	112.8	140.5

Source: Adapted from Marengo et al. 2006. *Environmental Science and Technology* 40: 272–280. With permission from the American Chemical Society.

PLS showing slightly greater predictive ability than PCR. For NO₂ emission data, the neural network's predictive ability was significantly lowered; however, it did prove superior to both PLS and PCR. The authors noted that the neural network model could thus not be effectively used for predicting NO₂ emissions using kiln conditions and the physicochemical properties of raw material and fuels as variables.

Given the "black box" nature of neural networks, the authors decided to calculate the first derivative of the best network to gain a better understanding of the influence of the input variables on the responses. A derivative is a measure of how a function changes as its input changes. Taking the first derivative allowed the authors to find the "true" influence of the input variables on SO₂ and dust responses, and thus used to help reduce the risk of exceeding standard limits through statistical control. Many of the original 19 were eliminated for both SO₂ (modified = 13 variables) and dust emission (modified = 12 variables), with those having positive constants used for developing reduced neural network models. Such refinement allowed for greater predictive ability and improved understanding of the chemical and physicochemical mechanisms for a more efficient, and less environmentally damaging, cement production process.

6.2.5 PARTITION COEFFICIENT PREDICTION

Konoz and Golmohammadi (2008) presented a detailed study on the development of a QSPR model for the prediction of air-to-blood partition coefficients of various volatile organic compounds (VOCs). Air-to-blood partition coefficients, described as

$$K_{blood} \frac{\text{Concentration}(\text{mol L}^{-1})\text{of compound in blood}}{\text{Concentration}(\text{mol L}^{-1})\text{of compound in air}} \qquad (6.1)$$

help express the pulmonary uptake and final distribution of VOCs in mammalian systems (Anderson, 1983). Partition coefficients between blood and air, and for that matter between tissue and blood, are prerequisite parameters for understanding the toxicokinetics of chemical compounds. As highlighted in the study, experimental determination of K_{blood} of various VOCs is expensive and requires a large amount of purified sample. Hence, the development of appropriate QSPR models from theoretically derived molecular descriptors that can calculate/predict molecular properties would be a useful undertaking, especially in regard to modeling the relationship between partition coefficients and the chemical properties of VOCs. All molecular descriptors were calculated using COmprehensive DEscriptors for Structural and Statistical Analysis (CODESSA) and DRAGON software based on molecular structural information and minimum energy molecular geometries, respectively.

Air-to-blood partition coefficients of 143 VOCs were taken from Abraham et al. (2005) and used as the study data set for all analyses. The neural network employed was a three-layer, fully connected, feedforward network with a sigmoidal transfer function. Descriptors selected by a genetic algorithm and specification of the constructed GA-MLR model are listed in Table 6.10. In the GA-MLR approach, a population of k strings is randomly chosen from the predictor matrix $\mathbf{X}$, with each string consisting of a row-vector with elements equal to the number of descriptive variables in the original data set (Hanrahan, 2009). The fitness of each string is evaluated with the use of the root-mean-square effort of prediction formula given by

$$\text{RMSEP} = \sqrt{\sum_{i=1}^{n_t} \frac{(\bar{y}_i - y_i)^2}{n_t}} \qquad (6.2)$$

where n_t = the number of objects in the test set, y_i = the known value of the property of interest for object i, and $\bar{y}$ = the value of the property of interest predicted by the model for object i. Recalling information on the evolutionary approach presented in

---

TABLE 6.10

Chosen Descriptors for Specification of Multiple Linear Regression Models

Descriptor	Notation	Coefficient
R maximal autocorrelation of lag 1 weighted by atomic Sanderson electronegativities	R1E+	4.456 (±0.917)
Electron density of the most negative atom	EDNA	0.343 (±0.059)
Maximum values of partial charge for C atoms	MXPCC	−3.341 (±1.976)
Surface-weighted CPSA	WNSA1	0.058 (±0.007)
Fractional CPSA	FNSA2	9.477 (±1.351)
Atomic-charge-weighted PPSA	PPSA3	0.316 (±0.052)
Constant		−2.941 (±0.377)

Source: From Konoz and Golmohammadi. 2008. *Analytical Chimica Acta* 619: 157–164. With permission from Elsevier.

TABLE 6.11

Architecture and Specification of the Optimized Neural Network Model

Number of nodes in the input layer	6
Number of nodes in the hidden layer	6
Number of nodes in the output layer	1
Weights learning rate	0.5
Bias learning rate	0.5
Momentum	0.4
Transfer function	Sigmoid

Source: From Konoz and Golmohammadi. 2008. *Analytical Chimica Acta* 619: 157–164. With permission from Elsevier.

Chapter 4, the strings with the highest fitness are selected and repeatedly subjected to crossovers and mutations until investigators are content with the model statistics.

Appendix III presents the 143-member data set and the corresponding observed neural network and MLR predicted values of air-to-blood partition coefficients of all molecules studied. Table 6.11 shows the architecture and specification of the optimized neural network. In comparison, the standard errors of training, test, and validation sets for the GA-MLR were 0.586, 0.350, and 0.376, respectively, which differ from values obtained from the optimized neural network at 0.095, 0.148, and 0.120, respectively. Overall, results obtained showed that the nonlinear neural network model proved valuable in accurately simulating the relationship between structural descriptors and air-to-blood partition coefficients of a variety of VOCs.

6.2.6 Neural Networks and the Evolution of Environmental Change (A Contribution by Kudłak et al.)

Human-induced environmental change is of regional and international concern. Although such change is not exclusive to the modern era, the twentieth century is in a unique period of human history given that it liberated the spirit and ingenuity of the people, while at the same time causing enormous changes in political thought and modifications to Earth's geospheres: the atmosphere, lithosphere, hydrosphere, biosphere, and pedosphere. The enormity of environmental change during this era strongly suggests that history, ecology, and technology are inextricably linked. Modern history, written as if the life-support systems of the planet were stable and present only in the background of human affairs, is not only incomplete but also misleading. Ecology and technology that neglects the complexity of social forces and dynamics of historical change are equally restricted. These are all fields of knowledge, and thus supremely integrative. If and when they are integrated, we will have a picture of our past that is more complete, more compelling, and more comprehensible. We will have a better idea of our present situation, and whether or not it qualifies as a predicament. And with that we will have a better idea of our possible future.

The relative importance of modeling tools in the assessment of environmental degradation and change cannot be overstressed. Costanza and Patten (1995) describes the latter concept in terms of characteristic elements, including a sustainable scale of the economy relative to its ecological life-support system, an equitable distribution of resources and opportunities between present and future generations, and an efficient allocation of resources that adequately accounts for natural capital. As will be evident in this section, neural networks are increasingly used in environmental risk assessment, and as tools for finding complex relationships between human activities and ecological processes. This methodology will be demonstrated in each of the following sections with the inclusion of a selected application.

6.2.6.1   Studies in the Lithosphere

Climate scientists have been reluctant to make a direct link between human-induced climate change and natural disasters, but they do predict that climate change will lead to more frequent extreme weather conditions. Consider the increased occurrence of typhoons that are striking countries such as Taiwan and bringing abundant rainfall inducing life-threatening landslides and creek debris flow. Debris flow in particular has been shown to cause serious loss of human lives and extensive property damage in this region. Realizing this, Chang (2007) developed neural network models to aid in the prediction of debris flow. An MLP using a back-propagation algorithm was constructed with seven input factors: (1) length of creek, (2) average slope, (3) effective watershed area, (4) shape coefficient, (5) median size of soil grain, (6) effective cumulative rainfall, and (7) effective rainfall intensity. An aggregate of 171 potential cases of debris flows collected in eastern Taiwan were fed into the model for training and testing purposes. The average ratio of successful model prediction reached 99.12%, which demonstrates that the optimized neural network model with seven significant factors can provide a highly stable and reliable result for the prediction of debris flows in hazard mitigation.

6.2.6.2   Studies in the Atmosphere

Atmospheric contaminants are a consequence of both natural (e.g., volcanic episodes) and anthropogenic (e.g., vehicular traffic and industrial) emission processes, chemical reactions, solar radiation, temperature, and related interactive processes. For example, tropospheric (ground-level) ozone is formed by the interaction of sunlight (particularly ultraviolet light) with hydrocarbons and nitrogen oxides (NO_x), which are emitted by automobiles, gasoline vapors, fossil fuel power plants, refineries, and other industrial processes. Recently, Spanish researchers used neural network models for estimating ambient ozone concentrations with surface meteorological and vehicle emission variables as predictors (Gomez-Sanchis et al., 2006). Three data sets (April of 1997, 1999, and 2000) from the Agriculture Training Centre of Carcaixent, Spain, were each split into two: a training set formed by the first 20 days of the month and a validation set formed by the remaining days, were used.

The authors evaluated several thousand MLP models. In order to select the most appropriate, they took into account the training mode (batch or online), the architecture (one or two hidden layers, with a number of hidden neurons in each one from 5 to 16), and the synaptic weight initialization (100 different initializations for each trained architecture). Models were finally selected by a cross-validation procedure. The best neural network models (in terms of the lowest MSE values) were formed by only one hidden layer with 7, 14, and 16 hidden neurons for years 1997, 1999, and 2000, respectively. An online training algorithm provided the best model for years 1997 and 1999, whereas a batch training algorithm realized the best results for year 2000. Two sensitivity analyses were performed: (1) an overall sensitivity analysis of input variables and (2) a sensitivity analysis dependent on the time of ozone concentration measurements. For instance, climatological variables directly related to temperature, transport elements, and NO_2 were found to be the most significant input factors concerning optimized models. The authors admit that while these variables represent an excellent starting point, models that take into account other variables (and other locations of study) must be considered for future study.

6.2.6.3 Studies in the Hydrosphere

Natural and seminatural (partially constructed) wetlands have been shown to contribute to sustainable water management activities in the 21st century. They perform a key ecological function in hydrological regulation and mass cycling, thus contributing to a variety of protection measures (Vymazal, 2009). Many communities around the globe are beginning to incorporate wetlands into watershed planning to help meet local water quality standards or permit regulations, generate revenue for the local economy, and protect and beautify wildlife habitats. The ability to assess the quality of outflow water, especially when water treatment monitoring and assessment are of concern, is of prime importance. A study by Lee and Scholz (2006) examined the goodness of applying K-nearest neighbors (KNNs), SVM, and SOMs to predict the outflow water quality of experimental constructed treatment wetlands by comparing the accuracy of these models. Additionally, this study described how machine learning can be effectively used in water treatment monitoring and assessment.

Experimental data were collected by monitoring the effluent concentrations of the 12 wetland filters, including biological oxygen demand (BOD) and suspended solids (SS) from September 9, 2002, to September 21, 2004. These data were stored in the database together with up to six input variables: turbidity (NTU), conductivity (μS), redox potential (mV), outflow water temperature (°C), dissolved oxygen, DO (mg L^{-1}), and pH. The corresponding output variables were BOD (mg L^{-1}) or SS (mg L^{-1}). The input variables were selected according to their goodness of correlation with both BOD and SS. In terms of predictive capabilities, SOM showed a better performance compared to both KNN and SVM. Furthermore, SOM had the capability of visualizing the relationship between complex biochemical variables present in the study. The authors did note that the SOM was more time intensive given its overall trial-and-error process when searching for the optimal map. In the end, the results proved that BOD and SS could be efficiently estimated by applying machine learning tools in real time. They were encouraged by the performance and recommend such techniques for future use in day-to-day process control activities.

6.2.6.4   Studies in the Biosphere

Our actions are also having considerable consequences in the biosphere, with habitat loss and fragmentation, and disturbance and introduction of alien species posing the greatest threats to biodiversity (Jordan et al., 2010). However, biodiversity management is complicated and often limited due to lack of information on species number and composition attributes. But as shown by Foody and Cutler (2006), remote sensing has considerable potential as a source of data on biodiversity at spatial and temporal scales appropriate for biodiversity management. When combined with neural networks, remote sensing has the ability to estimate biodiversity more fully. In this study, two neural network models were developed and implemented: (1) feedforward networks to estimate basic indices of biodiversity and (2) Kohonen networks to provide information on species composition. The test site was a 300 km^2 region of tropical forest surrounding the Danum Valley Field Centre in north eastern Borneo, Malaysia, with ground data acquired from 52 circular sample plots across a wide range of forest types.

   In terms of results, the authors reported that biodiversity indices of species richness and evenness derived from the remotely sensed data were well correlated with those derived from field survey. For example, the predicted tree species richness was significantly correlated with that observed in the field ($r = 0.69$, significant at the 95% level of confidence). Moreover, there was a high degree of correspondence (approximately 83%) between the partitioning of the outputs from Kohonen networks applied to tree species and remotely sensed data sets that indicated the potential to map species composition. Combining the outputs of the two sets of neural-network-based analyses enabled a map of biodiversity to be produced (e.g., Figure 6.11).

6.2.6.5   Environmental Risk Assessment

It is expected that neural networks will be regarded as meaningful tools for developing complex, multivariable risk assessment models. These models should be capable of fully simulating the source-pathway-receptor relationship to capture and understand the cause-effect linkage between concentration of chemicals and their influence on living organisms. Further development ought to incorporate an assortment of parameters now considered in separate models: characteristics relating to human activities, physiology (e.g., bodyweight, illnesses, and age), spatial and temporal effects, and the cumulative effects of different exposure pathways and multiple chemicals. Reflection on how to incorporate synergistic and/or antagonistic effects of a combination of chemicals is also important. However challenging the task, much time and attention should be dedicated to it as risk assessment is a fundamental issue allowing the comparison of the costs and benefits of economic growth.

6.3   CONCLUDING REMARKS

Reviewing the applications described in this chapter, one can conclude that neural network models have the potential to offer exceptional heuristic tools that allow insights into the effects of the numerous ecological mechanisms and human activities

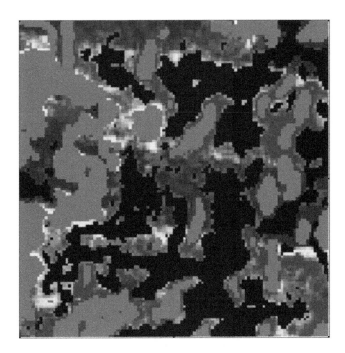

FIGURE 6.11 A representative map depicting the biodiversity information extracted from the neural-network-based analysis. This map indicates the variation in species richness for one of the four classes of species composition defined from the output of the Kohonen network applied to remotely sensed data. (From Foody and Cutler. 2006. *Ecological Modelling* 195: 37–42. With permission from Elsevier.)

that form the foundation of complex environmental systems. Nonetheless, the inevitable issue that remains is whether or not these models can truly represent natural systems and effectively model and predict natural and anthropogenic phenomena. With increasing application and study of environmental systems come a number of key considerations: complexity, variability, scale, purpose, model robustness, and sensitivity analysis. Such demands have led to the development of sophisticated models (including neural networks) utilizing today's computational capabilities and assessable resources. Although neural networks have enjoyed a long tradition in environmental investigations, there is a defined need to continually address both the power and limitations of such models, as well as their quantitative assessment. Only through continued use and refinement will their true success in environmental analysis be realized.

REFERENCES

Abraham, M.H., Ibrahim, A., and Acree, W.E. 2005. Chemical toxicity correlations for several fish species based on the Abraham solvation parameter model. *Chemical Research in Toxicology* 18: 904–911.

Anderson, M.E. 1983. *Flow Limited Clearance in Modeling of Inhalation Exposure to Vapors: Uptake Distribution, and Elimination*, Vol. II. CRC Press: Boca Raton, FL.

Barker, S.A. 1998. Matrix solid-phase dispersion. *LC-GC*. MAI Supplemental: S37–S40.

Barker, S.A. 2000. Applications of matrix solid-phase dispersion in food analysis. *Journal of Chromatography A* 880: 63–68.

Boszke, L. and Astel, A. 2009. Application of neural-based modeling in an assessment of pollution with mercury in the middle part of the Warta River. *Environmental Monitoring & Assessment* 152: 133–147.

Boti, V.I., Sakkas, V.A., and Albanis, T.A. 2009. An experimental design approach employing artificial neural networks for the determination of potential endocrine disruptors in food using matrix solid-phase dispersion. *Journal of Chromatography A*. 1216: 1296–1304.

Cakmakci, M. 2007. Adaptive neuro-fuzzy modelling of anaerobic digestion of primary sedimentation sludge. *Bioprocess and Biosystem Engineering* 30: 349–357.

Canpolat, B.R., Atimtay, A.T., Munlafalioglu, I., Kalafatoglu, E., and Ekinci, E. 2002. Emission factors of the cement industry in Turkey. *Water, Air and Soil Pollution* 138: 235–252.

Chandramouli, V., Brion, G., Neelakantan, T.R., and Lingireddy, S. 2007. Backfilling missing microbial concentrations in a riverine database using artificial neural networks. *Water Research* 41: 217–227.

Chang, T.-C. 2007. Risk degree of debris flow applying neural networks. *Natural Hazards* 42: 209–224.

Chau, K. 2006. A review of the integration of artificial intelligence into coastal modeling. *Journal of Environmental Management* 80: 47–57.

Chen, S.H., Jakeman, A.J., and Norton, J.P. 2008. Artificial intelligence techniques: An introduction to their use for modelling environmental systems. *Mathematics and Computers in Simulation* 78: 379–400.

Cherkassky, V., Krasnopolsky, V., Solomatine, D.P., and Valdes, J. 2006. Computational intelligence in earth sciences and environmental applications: Issues and challenges. *Neural Networks* 19: 113–121.

Costanza R., and Patten, B.C. 1995. Defining and predicting sustainability. *Ecological Economics* 15: 193–196.

Díaz-Robles, L.A., Ortega-Bravo, J.C., Fu, J.S., Reed, G.D., Chow, J.C., Watson, J.G. and Moncada, J.A. 2008. A Hybrid ARIMA and Artificial Neural Networks Model to Forecast Particulate Matter in Urban Areas: The Case of Temuco, Chile. *Atmospheric Environment* 42: 8331–8340.

Ferone, J.M., and Devito, K.J. 2004. Shallow groundwater-surface water interactions in pond-peatland complexes along a Boreal Plains topographic gradient. *Journal of Hydrology* 292: 75–95.

Ferré-Huguet, N., Nadal, M., Schuhmacher, M., and Domingo, J. 2006. Environmental impact and human health risks of polychlorinated dibenzo-*p*-dioxins and dibenzofurans in the vicinity of a new hazardous waste incinerator: A case study. *Environmental Science and Technology* 40: 61–66.

Foody, G.M., and Cutler, M.E.J. 2006. Mapping the species richness and composition of tropical forests from remotely sensed data with neural networks. *Ecological Modelling* 195: 37–42.

Gomez-Sanchis, J. Martin-Guerrero, J.D., Soria-Olivas, E., Vila-Frances, J., Carrasco, J.L., and dell Valle-Tascon, S. 2006. Neural networks for analyzing the relevance of input variables in the prediction of tropospheric ozone concentration. *Atmospheric Environment* 40: 6173–6180.

Hanrahan, G. 2009. *Environmental Chemometrics: Principles and Modern Applications*. CRC Press: Boca Raton, FL.

Hanrahan, G. Gledhill, M., House, W.A., and Worsfold, P.J. 2001. Phosphorus loading in the Frome catchment, UK: Seasonal refinement of the coefficient modeling approach. *Journal of Environmental Quality* 30: 1738–1746.

Ho, K.T., Kuhn, A., Pelletier, M., McGee, F., Burgess, R.M., and Serbst, J. 2000. Sediment toxicity assessment: Comparison of standard and new testing designs. *Archives of Environmental Contamination and Toxicology* 39: 462–468.

Jordan, S.J., Hayes, S.F., Yoskowitz, D., Smith, L.M., Summers, K., Russel, M., and Benson, W.H. 2010. Accounting for natural resources and environmental sustainability: Linking ecosystem services to human well-being. *Environmental Science and Technology* 44: 1530–1536.

Kasiri, M.B., Aleboyeh, H., and Aleboyeh, A. 2008. Modeling and optimization of heterogeneous photo-fenton process with response surface methodology and artificial neural networks. *Environmental Science and Technology* 42: 7970–7975.

Kim, C.Y., Bae, G.J., Hong, S.W., Park, C.H., Moon, H.K., and Shin, H.S. 2001. Neural network based prediction of ground surface settlements due to tunneling. *Computational Geotectonics* 28: 517–547.

Konoz, E., and Golmohammadi, H. 2008. Prediction of air-to-blood partition coefficients of volatile organic compounds using genetic algorithm and artificial neural network. *Analytical Chimica Acta* 619: 157–164.

Krasnopolsky, V.M., and Chevallier, F. 2003. Some neural network applications in environmental sciences. Part II: Advancing computational efficiency of environmental numerical models. *Neural Networks* 16: 335–348.

Krasnopolsky, V.M., and Schiller, H. 2003. Some neural network applications in environmental sciences. Part I: Forward and inverse problems in geophysical remote measurements. *Neural Networks* 16: 321–334.

Lau, K.-T., Guo, W., Kiernan, B., Slater, C., and Diamond, D. 2009. Non-linear carbon dioxide determination using infrared gas sensors and neural networks with Bayesian regularization. *Sensors and Actuators B: Chemical* 136: 242–47.

Lee, B.-H., and Scholz, M. 2006. A comparative study: Prediction of constructed treatment wetland performance with K-nearest neighbors and neural networks. *Water, Air and Soil Pollution* 174: 279–301.

Marcé, R., Comerma, M., García, J.C., and Armengol, J. 2004. A neuro-fuzzy modelling tool to estimate fluvial nutrient loads in watersheds under time-varying human impact. *Limnology and Oceanography: Methods* 2: 342–355.

Marengo, E., Bobba, M., Robotti, E., and Liparota, M.C. 2006. Modeling of the polluting emissions from a cement production plant by partial least-squares, principal component, and artificial neural networks. *Environmental Science and Technology* 40: 272–280.

May, R.J., Dandy, G.C., Maier, H.R., and Nixon, J.B. 2008. Application of partial mutual information variable selection to ANN forecasting of water quality in water distribution systems. *Environmental Modelling and Software* 23: 1289–1299.

May, R.J., Maier, H.R., and Dandy, G.C. 2009. Developing artificial neural networks for water quality modelling and analysis. In G. Hanrahan (Ed.), *Modelling of Pollutants in Complex Environmental Systems*. ILM Publications: St. Albans, U.K.

Muleta, M.K., and Nicklow, J.W. 2005. Decision support for watershed management using evolutionary algorithms. *Journal of Water Resources, Planning and Management* 131: 35–44.

Nagendra, S.M.S., and Khare, M. 2006. Artificial neural network approach for modelling nitrogen dioxide dispersion from vehicular exhaust emissions. *Ecological Modelling* 190: 99–115.

Nour, M.H., Smith, D.W., Gamal El-Din, M., and Prepas, E.E. 2006. The application of artificial neural networks to flow and phosphorus dynamics in small streams on the Boreal Plain, with emphasis on the role of wetlands. *Ecological Modelling* 191: 19–32.

Ooba, M., Hirano, T., Mogami, J.-I., Hirata, R., and Fujinuma, Y. 2006. Comparisons of gap-filling methods for carbon flux data set: A combination of a genetic algorithm and an artificial neural network. *Ecological Modelling* 198: 473–486.

Piuleac, C.G., Rodrigo, M.A., Cañizares, P., and Sáez, C. 2010. Ten steps modeling of electrolysis processes by using neural networks. *Environmental Modelling and Software* 25: 74–81.

Safe, S. 2004. Endocrine disruptors and human health: Is there a problem. *Toxicology* 205: 3–10.

Sarangi, P.K., Singh, N., Chauhan, R.K., and Singh, R. 2009. Short term load forecasting using artificial neural networks: A comparison with genetic algorithm implementation. *ARPN Journal of Engineering and Applied Sciences* 4: 88–93.

Sheinbaum, C., and Ozawa, L. 1998. Energy use and CO_2 emissions for Mexico's cement industry. *Energy* 23: 725–732.

Smith, A.G. 1991. Chlorinated Hydrocarbon Insecticides. In W.J. Hayes Jr. and E.R. Laws Jr. (Eds.), *Handbook of Pesticide Toxicology*. Academic Press: San Diego, CA, pp. 731–915.

Squire, S., and Ramsey, M.H. 2001. Inter-organisational sampling trials for the uncertainty estimation of landfill gas measurements. *Journal of Environmental Monitoring* 3: 288–294.

Tsakovski, S., Kudlak, B., Simeonov, V., Wolska, L., and Namiesnik, J. 2009. Ecotoxicity and chemical sediment data classification by the use of self-organising maps. *Analytica Chimica Acta* 631: 142–152.

Walmsley, A.D. 1997. Improved variable selection procedure for multivariate linear regression. *Analytical Chimica Acta* 354: 225–232.

Wierzba, S., and Nabrdalik, M. 2007. Biodegradation of cellulose in bottom sediments of Turawa Lake. *Physicochemical Problems of Mineral Processing* 41: 227–235.

Vymazal, J. 2009. The use of constructed wetlands with horizontal sub-surface flow for various types of wastewater. *Ecological Engineering* 35: 1–17.

Yi, Q.-X., Huang, J-F., Wang, F-M., Wang, X.-Z., and Liu, Z.-Y. 2007. Monitoring rice nitrogen status using hyperspectral reflectance and artificial neural network. *Environmental Science and Technology* 41: 6770–6775.

Appendix I: Review of Basic Matrix Notation and Operations

A matrix is a table of numbers consisting of n rows and m columns (i.e., an $n \times m$ matrix):

$$\mathbf{A} = \begin{pmatrix} a_{11} & a_{12} & a_{1m} \\ a_{21} & a_{22} & a_{2m} \\ \vdots & & \vdots \\ a_{n1} & a_{n2} & a_{nm} \end{pmatrix} \text{ or with real numbers: } \mathbf{A} = \begin{pmatrix} 1 & 4 & 6 \\ 2 & 9 & 7 \\ 3 & 5 & 8 \end{pmatrix}$$

where an individual element of $\mathbf{A}$ is a_{ij}. The first subscript in a matrix refers to the row and the second to the column. A square matrix consists of the same number of rows and columns (i.e., an $n \times n$ matrix). Note that matrix $\mathbf{A}$ above is square, but matrix $\mathbf{B}$ below is not:

$$\mathbf{B} = \begin{pmatrix} 1 & 4 & 6 \\ 2 & 9 & 7 \end{pmatrix}$$

A vector is a type of matrix that has only one row (i.e., a row vector) or one column (i.e., a column vector). Below, $\mathbf{a}$ is a column vector, while $\mathbf{b}$ is a row vector.

$$\mathbf{a} = \begin{pmatrix} 6 \\ 2 \end{pmatrix} \quad \mathbf{b} = (3 \quad 1 \quad 6)$$

A symmetric matrix is a square matrix in which $a_{ij} = a_{ji}$, for all i and j and is denoted as follows:

$$\mathbf{A} = \begin{pmatrix} 3 & 4 & 1 \\ 4 & 8 & 5 \\ 1 & 5 & 0 \end{pmatrix}$$

Do not confuse this with a diagonal matrix, which is a symmetric matrix where all the off-diagonal elements are 0:

$$\mathbf{D} = \begin{pmatrix} 3 & 0 & 0 \\ 0 & 1 & 0 \\ 0 & 0 & 5 \end{pmatrix}$$

An identity matrix is a diagonal matrix with 1's on the diagonal:

$$\mathbf{I} = \begin{pmatrix} 1 & 0 & 0 \\ 0 & 1 & 0 \\ 0 & 0 & 1 \end{pmatrix}$$

In a transposed matrix A', the rows and columns are interchanged as below. Compare that to the first matrix presented above.

$$\mathbf{A} = \begin{pmatrix} a_{11} & a_{21} & a_{n1} \\ a_{12} & a_{22} & a_{n2} \\ \vdots & & \vdots \\ a_{1m} & a_{2m} & a_{nm} \end{pmatrix} \text{ or with real numbers: } \mathbf{A} = \begin{pmatrix} 1 & 2 & 3 \\ 4 & 9 & 5 \\ 6 & 7 & 8 \end{pmatrix}$$

For matrix addition, each element of the first matrix is added to the corresponding element of the second to produce a result. Note that the two matrices must have the same number of rows and columns:

$$\begin{pmatrix} 0 & 2 & -1 \\ 3 & 1 & 3 \\ 1 & 6 & 0 \end{pmatrix} + \begin{pmatrix} 10 & -1 & 3 \\ 2 & 0 & 2 \\ 3 & 1 & 1 \end{pmatrix} = \begin{pmatrix} 10 & 1 & 2 \\ 5 & 1 & 5 \\ 4 & 7 & 1 \end{pmatrix}$$

Note that matrix subtraction works in the same way, except that elements are subtracted instead of added. Multiplication of an $n \times n$ matrix $\mathbf{A}$ and an $n \times n$ matrix $\mathbf{B}$ gives a result of $n \times n$ matrix $\mathbf{C}$:

$$\begin{pmatrix} 2 & 1 & 3 \\ -2 & 2 & 1 \end{pmatrix} \begin{pmatrix} 2 & 1 \\ 3 & 2 \\ -2 & 2 \end{pmatrix} = \begin{pmatrix} 1 & 10 \\ 0 & 4 \end{pmatrix}$$

Note that in multiplication, $\mathbf{A} \times \mathbf{B}$ does not generally equal $\mathbf{B} \times \mathbf{A}$. In other words, matrix multiplication is not commutative. Consider associative and distributive laws: $(\mathbf{A} \times \mathbf{B}) \times \mathbf{C} = \mathbf{A} \times (\mathbf{B} \times \mathbf{C})$; $\mathbf{A} \times (\mathbf{B} + \mathbf{C}) = \mathbf{A} \times \mathbf{B} + \mathbf{A} \times \mathbf{C}$ and $(\mathbf{B} + \mathbf{C}) \times \mathbf{A} = \mathbf{B} \times \mathbf{A} + \mathbf{C} \times \mathbf{A}$.

Typically, one takes the multiplication by an inverse matrix as the equivalent of matrix division. The inverse of a matrix is that matrix that when multiplied by the original matrix gives an identity ($\mathbf{I}$) matrix, the inverse being denoted by a superscripted −1:

$$\mathbf{A}^{-1}\mathbf{A} = \mathbf{A}\mathbf{A}^{-1} = \mathbf{I}$$

Note that to have an inverse, a matrix must be square. Consider the following matrix:

$$\mathbf{A} = \begin{pmatrix} a_{11} & a_{12} \\ a_{21} & a_{22} \end{pmatrix}$$

The inverse of this matrix exists if $a_{11}a_{22} - a_{12}a_{21} \neq 0$. If the inverse exists, it is given by

$$\mathbf{A} = \frac{1}{a_{11}a_{22} - a_{12}a_{21}} \begin{pmatrix} a_{22} & -a_{12} \\ -a_{21} & a_{11} \end{pmatrix}$$

For covariance and correlation matrices, an inverse will always exist, provided that there are more subjects than there are variables and that every variable has a variance greater than 0. The existence of the inverse is dependent on the determinant, a scalar-valued function of the matrix. For example:

$$\det \mathbf{A} = \begin{pmatrix} a_{11} & a_{12} \\ a_{21} & a_{22} \end{pmatrix} = a_{11}a_{22} - a_{12}a_{21}$$

For covariance and correlation matrices, the determinant is a number that often expresses the generalized variance of the matrix. Here, covariance matrices with small determinants denote variables that are highly correlated (e.g., factor analysis or regression analysis).

Defining the determinant can help in formalizing the general form of the inverse matrix:

$$\mathbf{A}^{-1} = \frac{1}{\det \mathbf{A}} adj\mathbf{A}$$

where $adj\mathbf{A}$ = the adjugate of $\mathbf{A}$. There are a couple of ways to compute an inverse matrix, the easiest typically being in the form of an augmented matrix ($\mathbf{A}|\mathbf{I}$) from

A and **In**, then utilizing Gaussian elimination to transform the left half into **I**. Once completed, the right half of the augmented matrix will be **A**⁻1. Additionally, one can compute the i,jth element of the inverse by using the general formula:

$$\mathbf{A}_{ji} = \frac{C_{ij}\mathbf{A}}{\det \mathbf{A}}$$

where $C_{ij} = i,j$th cofactor expansion of matrix **A**.

An orthogonal matrix has the general form $\mathbf{AA}^t = \mathbf{I}$. Thus, the inverse of an orthogonal matrix is simply its transpose. Orthogonal matrices are very important in factor analysis. Matrices of eigenvectors are orthogonal matrices. Note that only square matrices may be orthogonal matrices. As discussed earlier, eigenvalues and eigenvectors of a matrix play an important part in multivariate analysis. General concepts regarding eigenvalues and eigenvectors include the following:

1. Eigenvectors are scaled so that **A** is an orthogonal matrix.
2. An eigenvector of a linear transformation is a nonzero vector that is either left unaffected or simply multiplied by a scale factor after transformation.
3. The eigenvalue of a nonzero eigenvector is the scale factor by which it has been multiplied.
4. An eigenvalue reveals the proportion of total variability in a matrix associated with its corresponding eigenvector.
5. For a covariance matrix, the sum of the diagonal elements of the covariance matrix equals the sum of the eigenvalues.
6. For a correlation matrix, all the eigenvalues sum to n, the number of variables.
7. The decomposition of a matrix into relevant eigenvalues and eigenvectors rearranges the dimensions in an n-dimensional space so that all axes are perpendicular.

Appendix II: Cytochrome P450 (CYP450) Isoform Data Set Used in Michielan et al. (2009)

No.	Name	CYP1A2	CYP2C19	CYP2C8	CYP2C9	CYP2D6	CYP2E1	CYP3A4	Multilabel	Ref.
1	322	0	0	0	0	0	0	1	No	Metabolite Database
2	859	0	0	0	0	1	0	0	No	Metabolite Database
3	864	0	0	0	0	0	0	1	No	Metabolite Database
4	868	0	0	0	0	0	0	0	No	Metabolite Database
5	2689	0	0	0	1	0	0	0	No	Metabolite Database
6	3088	0	0	0	1	0	0	0	No	Metabolite Database
7	3179	0	0	0	1	0	0	0	No	Metabolite Database
8	3326	0	0	0	0	1	0	0	No	Metabolite Database
9	3676	0	0	0	0	1	0	0	No	Metabolite Database
10	3908	0	0	0	0	0	0	1	No	Metabolite Database
11	5018	0	0	0	0	0	0	1	No	Metabolite Database
12	5354	0	0	0	0	0	0	1	No	Metabolite Database
13	5476	0	0	0	0	1	0	0	No	Metabolite Database
14	5491	0	0	0	0	0	0	1	No	Metabolite Database
15	5716	0	0	0	0	1	0	0	No	Metabolite Database
16	5712	0	0	0	0	1	0	0	No	Metabolite Database
17	5726	0	0	0	0	1	0	0	No	Metabolite Database
18	6095	0	0	0	0	1	0	0	No	Metabolite Database
19	6228	1	0	0	0	0	0	0	No	Metabolite Database
20	6389	0	0	0	0	1	0	0	No	Metabolite Database
21	6609	0	0	0	0	1	0	0	No	Metabolite Database
22	6610	0	0	0	0	1	0	0	No	Metabolite Database
23	6634	0	0	0	0	1	0	0	No	Metabolite Database
24	6635	0	0	0	0	1	0	0	No	Metabolite Database
25	8606	0	0	0	0	1	0	0	No	Metabolite Database
26	8616	0	0	0	0	1	0	0	No	Metabolite Database
27	8641	0	1	0	0	0	0	0	No	Metabolite Database

#	ID									Metabolite Database
28	9093	0	0	0	0	0	0	1	No	Metabolite Database
29	9348	0	0	0	0	1	0	0	No	Metabolite Database
30	9862	0	0	0	0	1	0	0	No	Metabolite Database
31	9948	0	0	1	0	0	0	0	No	Metabolite Database
32	10011	1	0	0	0	0	0	0	No	Metabolite Database
33	10989	0	0	0	0	0	0	1	No	Metabolite Database
34	11120	0	0	0	1	0	0	0	No	Metabolite Database
35	11124	0	0	0	1	0	0	0	No	Metabolite Database
36	11295	0	0	0	1	0	0	0	No	Metabolite Database
37	11300	0	0	0	1	0	0	0	No	Metabolite Database
38	11441	0	1	0	0	1	0	0	No	Metabolite Database
39	11443	0	0	0	0	0	0	0	No	Metabolite Database
40	11447	0	1	0	0	0	0	0	No	Metabolite Database
41	11449	0	0	0	0	1	0	0	No	Metabolite Database
42	11750	1	0	0	0	0	0	0	No	Metabolite Database
43	12345	0	0	0	0	1	0	0	No	Metabolite Database
44	12451	0	0	0	0	1	0	0	No	Metabolite Database
45	13212	0	0	0	0	1	0	0	No	Metabolite Database
46	13414	0	0	0	0	1	0	0	No	Metabolite Database
47	13499	0	0	0	0	1	0	0	No	Metabolite Database
48	14334	0	0	0	1	0	0	0	No	Metabolite Database
49	14343	0	0	0	1	0	0	0	No	Metabolite Database
50	14467	0	1	0	0	0	0	0	No	Metabolite Database
51	14628	0	0	0	0	1	0	0	No	Metabolite Database
52	15061	0	0	0	0	1	1	0	No	Metabolite Database
53	15508	0	0	0	0	0	1	0	No	Metabolite Database
54	15852	0	0	0	0	0	0	0	No	Metabolite Database
55	16165	0	0	0	0	0	1	0	No	Metabolite Database

(Continued)

No.	Name	CYP1A2	CYP2C19	CYP2C8	CYP2C9	CYP2D6	CYP2E1	CYP3A4	Multilabel	Ref.
56	16396	0	0	0	0	0	1	0	No	Metabolite Database
57	16439	1	0	0	0	0	0	0	No	Metabolite Database
58	16457	0	0	0	0	0	1	0	No	Metabolite Database
59	16460	0	0	0	0	0	1	0	No	Metabolite Database
60	16462	0	0	0	0	0	1	0	No	Metabolite Database
61	16608	0	0	0	0	0	1	0	No	Metabolite Database
62	16611	0	0	0	0	0	1	0	No	Metabolite Database
63	16781	1	0	0	0	0	0	0	No	Metabolite Database
64	17178	0	0	0	0	0	0	1	No	Metabolite Database
65	17304	0	0	0	0	1	0	0	No	Metabolite Database
66	17308	0	0	0	0	1	0	0	No	Metabolite Database
67	18423	0	0	0	0	1	0	1	No	Metabolite Database
68	19636	0	0	0	0	0	0	0	No	Metabolite Database
69	19843	1	0	0	0	0	0	1	No	Metabolite Database
70	19860	1	0	0	0	0	0	0	No	Metabolite Database
71	20159	0	0	0	0	1	0	0	No	Metabolite Database
72	20440	0	0	0	0	1	0	0	No	Metabolite Database
73	21115	0	0	1	0	0	0	0	No	Metabolite Database
74	21125	0	0	0	1	0	0	0	No	Metabolite Database
75	21138	0	0	0	1	0	0	1	No	Metabolite Database
76	21342	0	0	0	0	0	0	0	No	Metabolite Database
77	21344	0	0	0	0	0	0	1	No	Metabolite Database
78	21491	0	0	0	0	1	0	0	No	Metabolite Database
79	22459	0	0	0	0	0	1	0	No	Metabolite Database
80	22636	1	0	0	0	0	0	0	No	Metabolite Database
81	23276	0	0	0	0	1	0	0	No	Metabolite Database
82	23281	0	0	0	0	1	0	0	No	Metabolite Database

#	ID									
83	23669	1	0	0	0	0	0	0	No	Metabolite Database
84	23677	1	0	0	0	0	0	0	No	Metabolite Database
85	24195	0	0	0	0	0	1	0	No	Metabolite Database
86	25165	0	0	0	0	1	0	0	No	Metabolite Database
87	25307	0	0	0	0	0	1	0	No	Metabolite Database
88	25318	0	0	0	0	0	0	1	No	Metabolite Database
89	25319	0	0	0	0	0	0	1	No	Metabolite Database
90	25414	0	0	0	0	0	0	1	No	Metabolite Database
91	25647	0	0	0	0	0	1	0	No	Metabolite Database
92	26299	1	0	0	0	0	0	0	No	Metabolite Database
93	27253	0	0	0	0	0	0	1	No	Metabolite Database
94	27906	0	0	0	0	0	0	1	No	Metabolite Database
95	27907	0	0	0	0	0	0	1	No	Metabolite Database
96	27909	0	0	0	0	1	0	1	No	Metabolite Database
97	29127	1	0	0	0	1	0	0	No	Metabolite Database
98	29870	0	0	0	0	0	0	0	No	Metabolite Database
99	29889	0	0	1	0	0	0	0	No	Metabolite Database
100	30029	0	0	0	0	0	0	0	No	Metabolite Database
101	30031	0	0	0	0	1	0	1	No	Metabolite Database
102	30034	0	0	0	0	0	0	1	No	Metabolite Database
103	30085	0	0	0	0	0	0	0	No	Metabolite Database
104	30095	0	0	0	0	0	0	1	No	Metabolite Database
105	30529	0	0	0	0	0	1	1	No	Metabolite Database
106	30531	0	0	0	0	0	0	1	No	Metabolite Database
107	31064	1	0	0	0	0	0	0	No	Metabolite Database
108	31537	1	0	0	0	0	0	0	No	Metabolite Database
109	31543	1	0	0	0	0	0	0	No	Metabolite Database
110	31893	0	0	0	0	0	0	1	No	Metabolite Database

(Continued)

No.	Name	CYP1A2	CYP2C19	CYP2C8	CYP2C9	CYP2D6	CYP2E1	CYP3A4	Multilabel	Ref.
111	34120	0	0	0	0	0	0	1	No	Metabolite Database
112	34454	1	0	0	0	0	0	0	No	Metabolite Database
113	34455	1	0	0	0	0	0	0	No	Metabolite Database
114	34457	1	0	0	0	0	0	0	No	Metabolite Database
115	34458	1	0	0	0	0	0	0	No	Metabolite Database
116	34475	0	0	0	0	0	0	1	No	Metabolite Database
117	34579	0	0	0	0	0	1	0	No	Metabolite Database
118	34690	0	0	0	0	0	1	0	No	Metabolite Database
119	35597	0	0	0	0	0	1	0	No	Metabolite Database
120	35600	0	0	0	0	0	1	0	No	Metabolite Database
121	36005	0	0	0	0	0	0	1	No	Metabolite Database
122	36731	0	0	0	0	0	1	0	No	Metabolite Database
123	37213	0	0	0	0	0	0	1	No	Metabolite Database
124	37474	0	0	0	0	0	0	1	No	Metabolite Database
125	37599	0	0	0	0	0	1	0	No	Metabolite Database
126	38395	1	0	0	0	0	0	0	No	Metabolite Database
127	38398	1	0	0	0	0	0	0	No	Metabolite Database
128	38402	0	1	0	0	0	0	0	No	Metabolite Database
129	38404	0	1	0	0	0	0	0	No	Metabolite Database
130	38407	0	1	0	0	0	0	0	No	Metabolite Database
131	38416	1	0	0	0	0	0	0	No	Metabolite Database
132	38528	1	0	0	0	0	0	0	No	Metabolite Database
133	38680	1	0	0	0	1	0	0	No	Metabolite Database
134	38744	0	0	0	0	0	0	1	No	Metabolite Database
135	39279	0	0	0	0	0	0	1	No	Metabolite Database
136	40334	0	0	0	0	0	0	1	No	Metabolite Database
137	40472	0	0	0	0	0	0	1	No	Metabolite Database

138	40931	0	1	0	0	0	0	0	No	Metabolite Database
139	40994	1	0	0	1	0	0	0	No	Metabolite Database
140	41354	0	0	0	0	0	0	1	No	Metabolite Database
141	41405	0	0	0	0	0	0	1	No	Metabolite Database
142	43025	0	0	0	0	0	0	1	No	Metabolite Database
143	43253	0	0	0	0	1	0	0	No	Metabolite Database
144	43851	0	0	0	0	0	0	1	No	Metabolite Database
145	44100	0	0	0	0	1	0	0	No	Metabolite Database
146	44150	1	0	0	0	0	1	0	No	Metabolite Database
147	44969	0	0	0	0	0	1	0	No	Metabolite Database
148	44972	0	0	0	0	0	0	0	No	Metabolite Database
149	44978	0	0	0	0	0	1	0	No	Metabolite Database
150	45411	0	0	0	0	1	0	0	No	Metabolite Database
151	46098	1	0	0	0	0	0	0	No	Metabolite Database
152	46231	0	0	0	0	0	0	0	No	Metabolite Database
153	46232	0	0	0	0	0	0	1	No	Metabolite Database
154	46346	1	0	0	0	0	0	1	No	Metabolite Database
155	46350	1	0	0	0	0	0	0	No	Metabolite Database
156	47019	0	0	0	0	1	0	0	No	Metabolite Database
157	47020	1	0	0	0	0	0	0	No	Metabolite Database
158	47041	0	0	0	0	1	0	0	No	Metabolite Database
159	47755	0	0	0	0	0	0	0	No	Metabolite Database
160	48285	0	0	0	0	0	0	1	No	Metabolite Database
161	48336	0	0	0	0	1	0	1	No	Metabolite Database
162	48337	0	0	0	0	1	0	0	No	Metabolite Database
163	48566	1	0	0	0	0	0	0	No	Metabolite Database
164	48657	0	0	0	0	0	0	1	No	Metabolite Database

(*Continued*)

No.	Name	CYP1A2	CYP2C19	CYP2C8	CYP2C9	CYP2D6	CYP2E1	CYP3A4	Multilabel	Ref.
165	48894	0	0	0	0	0	1	0	No	Metabolite Database
166	48895	0	0	0	0	0	1	0	No	Metabolite Database
167	49260	0	0	0	0	0	1	0	No	Metabolite Database
168	50413	0	0	0	1	0	0	0	No	Metabolite Database
169	50588	0	0	0	1	0	0	0	No	Metabolite Database
170	50826	0	0	0	0	0	0	1	No	Metabolite Database
171	51201	0	0	0	0	0	0	1	No	Metabolite Database
172	51488	0	0	0	0	0	0	1	No	Metabolite Database
173	51727	0	0	0	0	0	0	1	No	Metabolite Database
174	51728	0	0	0	0	0	0	1	No	Metabolite Database
175	51830	0	0	0	0	0	0	1	No	Metabolite Database
176	51831	0	0	0	0	0	0	1	No	Metabolite Database
177	52799	0	0	0	0	0	1	0	No	Metabolite Database
178	52801	0	0	0	0	0	1	0	No	Metabolite Database
179	53365	0	0	0	0	0	0	1	No	Metabolite Database
180	53367	0	0	0	1	0	0	0	No	Metabolite Database
181	55175	0	0	0	0	0	0	1	No	Metabolite Database
182	55955	0	0	0	1	0	0	0	No	Metabolite Database
183	57002	0	0	0	0	0	0	1	No	Metabolite Database
184	57004	0	0	0	0	0	0	1	No	Metabolite Database
185	57197	0	0	0	0	0	1	0	No	Metabolite Database
186	58162	0	0	0	0	0	0	1	No	Metabolite Database
187	58164	0	0	0	0	0	0	1	No	Metabolite Database
188	58173	0	0	0	0	0	0	1	No	Metabolite Database
189	58176	0	0	0	0	0	0	1	No	Metabolite Database
190	59883	0	0	0	0	0	1	0	No	Metabolite Database
191	59887	0	0	0	0	0	1	0	No	Metabolite Database

#	ID	C1	C2	C3	C4	C5	C6	C7		Source
192	59889	0	1	0	0	0	0	0	No	Metabolite Database
193	60153	1	0	0	0	0	0	0	No	Metabolite Database
194	60586	1	0	0	0	0	0	0	No	Metabolite Database
195	60616	0	0	1	0	0	0	0	No	Metabolite Database
196	60821	0	0	1	0	0	0	0	No	Metabolite Database
197	60822	0	0	1	0	0	0	0	No	Metabolite Database
198	60823	0	0	1	1	0	0	0	No	Metabolite Database
199	60824	0	0	0	0	0	0	0	No	Metabolite Database
200	61327	0	0	1	0	0	0	0	No	Metabolite Database
201	61469	0	0	0	0	0	0	1	No	Metabolite Database
202	61470	0	0	0	0	0	0	1	No	Metabolite Database
203	61472	0	0	0	0	0	0	1	No	Metabolite Database
204	61650	0	0	0	0	0	0	1	No	Metabolite Database
205	61815	0	0	1	0	0	0	0	No	Metabolite Database
206	61816	0	0	1	0	0	0	0	No	Metabolite Database
207	62027	1	0	0	0	0	0	0	No	Metabolite Database
208	62126	1	0	0	0	0	0	0	No	Metabolite Database
209	62636	0	0	1	0	0	0	0	No	Metabolite Database
210	62744	0	0	1	0	0	0	0	No	Metabolite Database
211	63097	0	1	0	0	0	0	0	No	Metabolite Database
212	63098	0	1	0	0	0	0	0	No	Metabolite Database
213	63099	0	1	0	0	0	0	0	No	Metabolite Database
214	63100	0	1	0	0	0	0	0	No	Metabolite Database
215	63101	0	1	0	0	0	0	0	No	Metabolite Database
216	63102	0	1	0	0	0	0	0	No	Metabolite Database
217	63104	0	1	0	0	0	0	0	No	Metabolite Database
218	63115	1	0	0	0	0	0	0	No	Metabolite Database
219	63353	1	0	0	0	0	0	0	No	Metabolite Database

(Continued)

No.	Name	CYP1A2	CYP2C19	CYP2C8	CYP2C9	CYP2D6	CYP2E1	CYP3A4	Multilabel	Ref.
220	63824	0	0	0	0	0	0	1	No	Metabolite Database
221	63829	0	0	0	0	0	0	1	No	Metabolite Database
222	64618	0	0	0	0	0	0	1	No	Metabolite Database
223	64627	1	0	0	0	0	0	0	No	Metabolite Database
224	64819	0	0	0	0	0	0	1	No	Metabolite Database
225	64859	0	0	0	0	0	1	0	No	Metabolite Database
226	64929	0	0	0	0	0	0	1	No	Metabolite Database
227	65245	0	0	0	0	0	0	1	No	Metabolite Database
228	66235	0	0	0	0	0	0	1	No	Metabolite Database
229	66260	0	0	0	0	0	0	1	No	Metabolite Database
230	66337	0	0	0	0	0	0	1	No	Metabolite Database
231	66427	0	0	0	0	0	1	0	No	Metabolite Database
232	66478	0	0	0	0	0	0	1	No	Metabolite Database
233	67107	0	0	0	0	0	0	1	No	Metabolite Database
234	67111	0	0	0	0	0	0	1	No	Metabolite Database
235	67517	0	0	0	0	0	0	1	No	Metabolite Database
236	67596	0	0	0	0	0	0	1	No	Metabolite Database
237	68665	0	0	0	0	0	0	1	No	Metabolite Database
238	68790	0	0	1	0	0	0	0	No	Metabolite Database
239	68791	0	0	1	0	0	0	0	No	Metabolite Database
240	69572	0	0	0	1	0	0	0	No	Metabolite Database
241	69573	0	0	0	1	0	0	0	No	Metabolite Database
242	69575	0	0	0	1	0	0	0	No	Metabolite Database
243	69578	0	0	0	1	0	0	0	No	Metabolite Database
244	69579	0	0	0	1	0	0	0	No	Metabolite Database
245	69585	0	0	0	1	0	0	0	No	Metabolite Database
246	69734	0	0	0	0	0	0	1	No	Metabolite Database

#	ID									Flag	Reference
247	69807	0	0	1	0	0	0	0		No	Metabolite Database
248	70401	0	0	0	0	1	0	0		No	Metabolite Database
249	70678	0	0	0	0	0	0	1		No	Metabolite Database
250	70714	0	0	0	0	1	0	0		No	Metabolite Database
251	71062	0	0	0	0	1	0	0		No	Metabolite Database
252	71072	0	0	0	0	1	0	0		No	Metabolite Database
253	71098	1	0	0	0	0	0	0		No	Metabolite Database
254	71099	1	0	0	0	0	0	0		No	Metabolite Database
255	71410	0	0	0	0	0	0	1		No	Metabolite Database
256	72142	0	0	0	1	0	0	1		No	Metabolite Database
257	72588	0	0	0	0	0	0	0		No	Metabolite Database
258	72922	0	0	0	0	1	0	0		No	Metabolite Database
259	73006	1	0	0	0	0	0	1		No	Metabolite Database
260	73066	0	0	0	0	0	0	1		No	Metabolite Database
261	73076	0	0	0	0	1	0	0		No	Metabolite Database
262	73184	0	0	0	0	0	0	1		No	Metabolite Database
263	73186	0	0	0	0	0	0	1		No	Metabolite Database
264	73201	0	0	0	0	0	0	1		No	Metabolite Database
265	73211	0	0	0	0	0	0	1		No	Metabolite Database
266	72212	0	0	0	0	0	0	1		No	Metabolite Database
267	73334	0	0	0	0	0	0	1		No	Metabolite Database
268	Aceclofenac	0	0	1	0	0	0	0		No	Bonnabry (2001)
269	Acenocoumarol	0	0	1	0	1	0	0		Yes	Bonnabry (2001)
270	Acetaminophen	1	0	0	0	0	0	0		No	Block (2008)
271	Alfentanil	0	0	0	0	0	0	1		No	Block (2008)
272	Alfuzosin	0	0	0	0	0	0	1		No	Block (2008)
273	Almotriptan	0	0	0	0	0	0	1		No	Block (2008)
274	Alpidem	0	0	0	0	0	0	1		No	Manga (2005)

(Continued)

No.	Name	CYP1A2	CYP2C19	CYP2C8	CYP2C9	CYP2D6	CYP2E1	CYP3A4	Multilabel	Ref.
275	Alprazolam	0	0	0	0	0	0	1	No	Block (2008)
276	Alprenolol	0	0	0	0	1	0	0	No	Drug Interaction
277	Amiflamine	0	0	0	0	1	0	0	No	Manga (2005)
278	Amiodarone	0	0	1	0	0	0	1	Yes	Block (2008)
279	Amitriptyline	1	1	0	1	1	0	1	Yes	Block (2008)
280	Amlodipine	0	0	0	0	0	0	1	No	Block (2008)
281	Amphetamine	0	0	0	0	1	0	0	No	Block (2008)
282	Amprenavir	0	0	0	0	0	0	1	No	Block (2008)
283	Aprepitant	0	0	0	0	0	0	1	No	Block (2008)
284	Aripiprazole	0	0	0	0	1	0	1	Yes	Drug Interaction
285	Astemizole	0	0	0	0	0	0	1	No	Block (2008)
286	Atazanavir	0	0	0	0	0	0	1	No	Block (2008)
287	Atomexetine	0	0	0	0	1	0	0	No	Block (2008)
288	Atorvastatin	0	0	0	0	0	0	1	No	Block (2008)
289	Azatadine	0	0	0	0	0	0	1	No	Block (2008)
290	Beclomethasone	0	0	0	0	0	0	1	No	Block (2008)
291	Benzphetamine	0	0	1	0	0	0	0	No	Block (2008)
292	Bepridil	0	0	0	0	0	0	1	No	Block (2008)
293	Bexarotene	0	0	0	0	1	0	0	No	Block (2008)
294	Bisoprolol	0	0	0	0	0	0	1	No	Block (2008)
295	Bromocriptine	0	0	0	0	0	0	1	No	Block (2008)
296	Budesonide	0	0	0	0	0	0	1	No	Block (2008)
297	Bufuralol	0	0	0	0	1	0	0	No	Manga (2005)
298	Buspirone	0	0	0	0	0	0	1	No	Block (2008)
299	Busulfan	0	0	0	0	0	0	1	No	Block (2008)
300	Caffeine	1	1	0	0	0	0	1	Yes	Block (2008)
301	Carbamazepine	0	0	1	0	0	0	1	Yes	Block (2008)

302	Carisoprodol	0	1	0	0	0	0	No	Block (2008)
303	Carmustine	0	0	0	0	0	1	No	Bonnabry (2001)
304	Carvedilol	0	0	0	1	0	0	Yes	Block (2008)
305	Celecoxib	0	0	0	1	0	0	No	Block (2008)
306	Cerivastatin	0	0	1	0	0	1	Yes	Drug Interaction
307	Cevimeline	0	0	0	0	0	1	Yes	Block (2008)
308	Chlordiazepoxide	1	0	0	0	0	0	No	Block (2008)
309	Chlorpheniramine	0	0	0	0	0	1	No	Block (2008)
310	Chlorpromazine	0	0	0	1	0	0	No	Block (2008)
311	Chlorpropamide	0	0	0	1	0	0	No	Block (2008)
312	Chlorzoxazone	0	0	0	0	1	0	No	Bonnabry (2001)
313	Cilostazol	0	1	0	0	0	1	Yes	Block (2008)
314	Cinacalcet	1	0	0	1	0	1	Yes	Block (2008)
315	Cinnarizine	0	0	0	1	0	0	No	Manga (2005)
316	Cisapride	0	0	0	0	0	1	No	Block (2008)
317	Citalopram	0	1	0	0	0	1	Yes	Block (2008)
318	Clarithromycin	0	0	0	0	0	1	No	Block (2008)
319	Clindamycin	0	0	0	0	0	1	No	Block (2008)
320	Clomipramine	1	1	1	1	1	1	Yes	Block (2008)
321	Clonazepam	0	0	0	0	0	1	No	Block (2008)
322	Clopidogrel	1	0	0	0	0	1	Yes	Block (2008)
323	Clozapine	1	0	0	1	0	1	Yes	Block (2008)
324	Cocaine	0	0	0	0	0	1	No	Block (2008)
325	Codeine	0	0	0	1	1	0	No	Block (2008)
326	Colchicine	0	0	0	0	0	1	No	Block (2008)
327	Cortisol	0	0	0	0	0	1	No	Block (2008)
328	Cyclobenzaprine	1	0	0	1	1	1	Yes	Block (2008)
329	Cyclophosphamide	0	1	0	0	0	1	Yes	Block (2008)

(*Continued*)

No.	Name	CYP1A2	CYP2C19	CYP2C8	CYP2C9	CYP2D6	CYP2E1	CYP3A4	Multilabel	Ref.
330	Cyclosporin	0	0	0	0	0	0	1	No	Bonnabry (2001)
331	Dapsone	0	0	0	0	0	1	1	Yes	Block (2008)
332	Darifenacin	0	0	0	0	1	0	1	Yes	Block (2008)
333	Debrisoquine	0	0	0	0	1	0	0	No	Drug Interaction
334	Delavirdine	0	0	0	0	0	0	1	No	Block (2008)
335	Deprenyl	0	0	0	0	1	0	0	No	Manga (2005)
336	Desipramine	1	0	0	0	1	0	0	Yes	Block (2008)
337	Desogestrel	0	1	0	0	0	0	1	Yes	Block (2008)
338	Dexamethasone	0	0	0	1	0	0	1	No	Block (2008)
339	Dexfenfluramine	0	0	0	0	1	0	0	No	Block (2008)
340	Dextromethorphan	0	0	0	0	1	0	1	Yes	Block (2008)
341	Diazepam	1	1	0	1	0	0	1	Yes	Block (2008)
342	Diclofenac	0	0	0	1	0	0	0	No	Block (2008)
343	Dihydrocodeine	0	0	0	0	1	0	1	Yes	Bonnabry (2001)
344	Dihydroergotamine	0	0	0	0	0	0	1	No	Block (2008)
345	Diltiazem	0	0	0	0	0	0	1	No	Block (2008)
346	Disopyramide	0	0	1	0	0	0	1	No	Block (2008)
347	Docetaxel	0	0	0	0	0	0	1	Yes	Block (2008)
348	Dofetilide	0	0	0	0	0	0	1	No	Block (2008)
349	Dolasetron	0	0	0	0	0	0	1	Yes	Block (2008)
350	Domperidone	0	0	0	0	0	0	1	No	Drug Interaction
351	Donezepil	0	0	0	0	1	0	1	Yes	Block (2008)
352	Doxepin	0	0	0	0	1	0	0	No	Block (2008)
353	Doxorubicin	0	0	0	0	1	0	0	No	Block (2008)
354	Dronabinol	0	0	0	1	0	0	1	Yes	Block (2008)
355	Duloxetine	1	0	0	0	1	0	0	Yes	Block (2008)
356	Dutasteride	0	0	0	0	0	0	1	No	Block (2008)

No.	Drug								Yes/No	Reference
357	Ebastine	0	0	0	0	0	0	1	No	Manga (2005)
358	Efavirenz	0	0	0	0	0	0	1	No	Block (2008)
359	Enalapril	0	0	0	0	0	0	1	No	Manga (2005)
360	Encainide	0	0	0	0	1	0	0	No	Block (2008)
361	Enflurane	0	0	0	0	0	1	0	No	Bonnabry (2001)
362	Eplerenone	0	0	0	0	0	0	1	No	Block (2008)
363	Ergotamine	1	0	0	0	0	0	1	No	Block (2008)
364	Erlotinib	0	0	0	0	0	0	1	Yes	Block (2008)
365	Erythromycin	1	0	0	0	0	0	1	No	Block (2008)
366	Estradiol	0	0	0	0	0	0	0	No	Block (2008)
367	Eszopidone	0	0	0	0	0	1	1	Yes	Block (2008)
368	Ethanol	0	0	0	0	0	1	0	No	Bonnabry (2001)
369	Ethinylestradiol	0	0	0	0	0	0	1	No	Block (2008)
370	Ethosuximide	0	0	0	0	0	0	1	No	Block (2008)
371	Ethylmorphine	0	0	0	0	1	1	1	Yes	Bonnabry (2001)
372	Etonogestrel	0	0	0	0	0	0	1	No	Block (2008)
373	Etoposide	0	0	0	0	0	0	1	No	Block (2008)
374	Exemestane	0	0	0	0	0	0	1	No	Block (2008)
375	Felodipine	0	0	0	0	0	0	1	No	Block (2008)
376	Fenfluramine	0	0	0	0	1	0	0	No	Block (2008)
377	Fentanyl	0	0	0	0	1	1	1	Yes	Block (2008)
378	Fexofenadine	0	0	0	0	0	0	1	No	Block (2008)
379	Finasteride	0	0	0	0	0	0	1	No	Block (2008)
380	Flecainide	0	0	0	0	1	0	0	No	Block (2008)
381	Fluconazole	0	0	0	0	0	0	1	No	Manga (2005)
382	Flunarizine	1	0	0	0	1	1	0	No	Manga (2005)
383	Flunitrazepam	0	0	0	0	0	0	1	Yes	Bonnabry (2001)
384	Fluoxetine	0	0	1	1	1	0	0	Yes	Block (2008)

(Continued)

No.	Name	CYP1A2	CYP2C19	CYP2C8	CYP2C9	CYP2D6	CYP2E1	CYP3A4	Multilabel	Ref.
385	Fluphenazine	0	0	0	0	1	0	0	No	Block (2008)
386	Flurbiprofen	0	0	0	1	0	0	0	No	Block (2008)
387	Flutamide	1	0	0	0	0	0	1	Yes	Block (2008)
388	Fluticasone	0	0	0	0	0	0	1	No	Block (2008)
389	Fluvastatin	0	0	1	1	0	0	0	Yes	Block (2008)
390	Fluvoxamine	1	0	0	0	1	0	0	Yes	Block (2008)
391	Formoterol	0	1	0	1	1	0	0	Yes	Block (2008)
392	Fulvestrant	0	0	0	0	0	0	1	No	Block (2008)
393	Galantamine	0	0	0	0	1	0	1	Yes	Block (2008)
394	Glimepiride	0	0	0	1	0	0	0	No	Block (2008)
395	Glipizide	0	0	0	1	0	0	0	No	Block (2008)
396	Glyburide	0	0	0	1	0	0	0	No	Block (2008)
397	Granisetron	0	0	0	0	0	0	1	No	Bonnabry (2001)
398	Haloperidol	1	0	0	0	1	1	1	Yes	Block (2008)
399	Halothane	0	0	0	0	0	1	0	No	Bonnabry (2001)
400	Hexobarbital	0	1	0	0	0	1	0	Yes	Block (2008)
401	Hydrocodone	0	0	0	0	1	0	1	Yes	Block (2008)
402	Ibuprofen	0	0	1	1	0	0	0	No	Block (2008)
403	Ifosfamide	0	0	0	0	0	0	1	No	Block (2008)
404	Imatinib	0	0	0	0	0	0	1	No	Block (2008)
405	Imipramine	1	1	0	1	1	0	1	Yes	Block (2008)
406	Indinavir	0	0	0	0	0	0	1	No	Block (2008)
407	Indomethacin	0	1	0	1	0	0	0	Yes	Block (2008)
408	Irbesartan	0	0	0	1	0	0	0	No	Block (2008)
409	Irinotecan	0	0	0	0	0	0	1	No	Drug Interaction
410	Isoflurane	0	0	0	0	0	1	0	No	Bonnabry (2001)
411	Isotretinoin	0	0	1	0	0	0	0	No	Block (2008)

412	Isradipine	0	0	0	0	1	No	Block (2008)
413	Itraconazole	0	0	0	0	1	No	Block (2008)
414	Ketoconazole	0	0	0	0	1	No	Block (2008)
415	Lansoprazole	0	1	0	0	1	Yes	Block (2008)
416	Lercanidipine	0	0	0	0	1	No	Drug Interaction
417	Letrozole	0	0	0	0	1	No	Block (2008)
418	Levobupivacaine	1	0	0	0	1	Yes	Block (2008)
419	Levonorgestrel	0	0	0	0	1	No	Manga (2005)
420	Lidocaine	0	0	0	1	1	Yes	Block (2008)
421	Lisuride	0	0	0	0	1	No	Manga (2005)
422	Lobeline	0	0	0	1	0	No	Mango (2005)
423	Lopinavir	0	0	0	0	1	No	Block (2008)
424	Loratadine	0	0	0	0	1	No	Block (2008)
425	Lornoxicam	0	0	1	0	0	No	Manga (2008)
426	Losartan	0	0	1	1	1	Yes	Block (2008)
427	Lovastatin	0	0	0	0	1	No	Block (2008)
428	Maprotiline	0	0	0	1	0	No	Block (2008)
429	Maraviroc	0	0	0	0	1	No	Block (2008)
430	Medroxyprogesterone	0	0	0	1	1	No	Block (2008)
431	Mefenamic acid	1	0	1	0	0	No	Block (2008)
432	Meloxicam	0	0	1	0	0	No	Block (2008)
433	Meperidine	0	0	0	1	0	No	Block (2008)
434	Mephobarbital	0	1	0	0	0	No	Block (2008)
435	Methadone	1	0	0	1	1	Yes	Block (2008)
436	Methamphetamine	0	0	0	1	0	No	Block (2008)
437	Methoxyamphetamine	0	0	0	1	0	No	Block (2008)
438	Methoxyflurane	0	0	0	0	0	No	Bonnabry (2001)
439	Methoxyphenamine	0	0	0	1	0	No	Manga (2005)

(Continued)

No.	Name	CYP1A2	CYP2C19	CYP2C8	CYP2C9	CYP2D6	CYP2E1	CYP3A4	Multilabel	Ref.
440	Methylprednisolone	0	0	0	0	0	0	1	No	Block (2008)
441	Metoclopramide	0	0	0	0	1	0	0	No	Drug Interaction
442	Metoprolol	0	0	0	0	1	0	0	No	Block (2008)
443	Mexiletine	1	0	0	0	1	0	0	Yes	Block (2008)
444	Mianserin	1	0	0	0	1	0	1	Yes	Block (2008)
445	Miconazole	0	0	0	0	0	0	1	No	Block (2008)
446	Midazolam	0	0	0	0	0	0	1	No	Block (2008)
447	Mifepristone	0	0	0	0	0	0	1	No	Block (2008)
448	Minaprine	0	0	0	0	1	0	0	No	Drug Interaction
449	Mirtazapine	1	0	0	0	1	0	1	Yes	Block (2008)
450	Moclobemide	0	1	0	0	0	0	0	No	Block (2008)
451	Modafinil	0	0	0	0	0	0	1	No	Block (2008)
452	Mometasone	0	0	0	0	0	0	1	No	Block (2008)
453	Montelukast	0	0	0	1	0	0	1	Yes	Block (2008)
454	Morphine	0	0	0	0	1	0	0	No	Block (2008)
455	Naproxen	1	0	0	1	0	0	0	Yes	Block (2008)
456	Nateglinide	0	0	0	1	0	0	1	Yes	Block (2008)
457	Nefazodone	0	0	0	0	0	0	1	No	Block (2008)
458	Nelfinavir	0	1	0	0	0	0	1	Yes	Block (2008)
459	Nevirapine	0	0	0	0	0	0	1	No	Block (2008)
460	Nicardipine	0	0	0	0	0	0	1	No	Block (2008)
461	Nifedipine	0	0	0	0	0	0	1	No	Block (2008)
462	Nilutamide	0	1	0	0	0	0	0	No	Block (2008)
463	Nimodipine	0	0	0	0	0	0	1	No	Block (2008)
464	Nisoldipine	0	0	0	0	0	0	1	No	Block (2008)
465	Nitrendipine	0	0	0	0	0	0	1	No	Block (2008)
466	Norethisterone	0	0	0	0	0	0	1	No	Block (2008)

#	Drug								Yes/No	Reference
467	Nortriptyline	1	0	0	0	1	0	0	Yes	Block (2008)
468	Olanzapine	1	0	0	0	1	0	0	Yes	Block (2008)
469	Omeprazole	0	1	0	1	0	1	1	Yes	Block (2008)
470	Ondansetron	1	0	0	0	1	0	1	Yes	Block (2008)
471	Oxybutynin	0	0	0	0	0	0	1	No	Block (2008)
472	Oxycodone	0	0	1	0	1	0	0	No	Block (2008)
473	Paclitaxel	0	0	0	1	0	1	1	Yes	Block (2008)
474	Pantoprazole	0	1	0	0	0	0	1	Yes	Block (2008)
475	Paroxetine	0	0	0	0	1	0	0	No	Block (2008)
476	Pentamidine	0	1	0	0	1	0	0	No	Block (2008)
477	Perhexiline	0	0	0	0	0	0	0	No	Drug Interaction
478	Perphenazine	0	0	0	0	1	0	0	No	Block (2008)
479	Phenformin	0	0	0	0	0	0	0	No	Drug Interaction
480	Phenylbutazone	0	1	1	0	0	0	0	No	Manga (2005)
481	Phenytoin	0	1	1	0	0	1	0	Yes	Block (2008)
482	Pimozide	0	0	0	0	0	0	1	No	Block (2008)
483	Pindolol	0	0	0	1	0	0	0	No	Block (2008)
484	Pioglitazone	0	1	1	0	0	1	0	Yes	Block (2008)
485	Piroxicam	0	0	1	0	0	0	0	No	Block (2008)
486	Prednisolone	0	0	0	0	0	0	1	No	Block (2008)
487	Prednisone	0	0	0	0	0	0	1	No	Block (2008)
488	Progesterone	0	0	0	1	1	0	1	Yes	Block (2008)
489	Proguanil	0	0	0	1	1	0	0	No	Block (2008)
490	Promethazine	0	1	0	0	0	0	0	No	Block (2008)
491	Propafenone	1	1	0	0	0	0	1	Yes	Block (2008)
492	Propoxyphene	0	1	0	0	0	0	1	No	Block (2008)
493	Propranolol	1	1	0	1	1	0	1	Yes	Block (2008)
494	Quercetin	0	0	0	0	0	0	1	No	Manga (2005)

(Continued)

No.	Name	CYP1A2	CYP2C19	CYP2C8	CYP2C9	CYP2D6	CYP2E1	CYP3A4	Multilabel	Ref.
495	Quetiapine	0	0	0	0	1	0	1	Yes	Block (2008)
496	Quinidine	0	0	0	0	0	0	1	No	Block (2008)
497	Quinine	0	0	0	0	0	0	1	No	Block (2008)
498	Rabeprazole	0	1	0	0	0	0	1	Yes	Block (2008)
499	Ramelteon	1	0	0	1	0	0	1	Yes	Block (2008)
500	Ranolzaine	0	0	0	0	1	0	1	Yes	Block (2008)
501	Remoxipride	0	0	0	0	1	0	0	No	Manga (2005)
502	Repaglinide	0	0	1	0	0	0	1	Yes	Block (2008)
503	Retinoic acid	0	0	1	0	0	0	0	No	Block (2008)
504	Retinol	0	0	1	0	0	0	0	No	Block (2008)
505	Rifabutin	0	0	0	0	0	0	1	No	Block (2008)
506	Rifampicin	0	0	0	0	0	0	1	No	Block (2008)
507	Riluzole	1	0	0	0	0	0	0	No	Block (2008)
508	Risperidone	0	0	0	0	1	0	0	No	Block (2008)
509	Ritnavir	0	0	0	0	1	0	1	Yes	Block (2008)
510	Ropirinole	1	0	0	0	0	0	0	No	Block (2008)
511	Ropivacaine	1	0	0	0	0	0	0	No	Block (2008)
512	Rosiglitazone	0	0	1	1	0	0	0	Yes	Block (2008)
513	Salmeterol	0	0	0	0	0	0	1	No	Block (2008)
514	Saquinavir	0	0	0	0	0	0	1	No	Block (2008)
515	Sertindole	0	0	0	0	0	0	1	No	Manga (2005)
516	Sertraline	0	0	0	0	0	0	1	No	Manga (2005)
517	Sevoflurane	0	0	0	0	0	1	0	No	Bonnabry (2001)
518	Sibutramine	0	0	0	0	0	0	1	No	Block (2008)
519	Sildenafil	0	0	0	1	0	0	1	Yes	Block (2008)
520	Simvastatin	0	0	0	0	0	0	1	No	Block (2008)
521	Sirolimus	0	0	0	0	0	0	1	No	Block (2008)

522	Solifenacin	0	0	0	0	0	0	No	Block (2008)
523	Sorafenib	0	0	0	0	0	1	No	Block (2008)
524	Sparteine	0	0	0	0	1	0	No	Manga (2005)
525	Sufentanil	0	0	0	0	0	1	No	Bonnabry (2001)
526	Sulfamethizole	0	0	0	1	0	0	No	Manga (2005)
527	Sulfamethoxazole	0	0	0	1	0	0	No	Block (2008)
528	Sulfidimidine	0	0	0	0	0	1	No	Manga (2005)
529	Sunitinib	0	0	0	0	0	1	No	Block (2008)
530	Suprofen	0	0	0	1	0	0	No	Drug Interaction
531	Tacrine	1	0	0	0	0	0	No	Block (2008)
532	Tacrolimus	0	0	0	0	0	1	No	Block (2008)
533	Tadalafil	0	0	0	0	0	1	No	Block (2008)
534	Tamoxifen	0	0	0	1	1	1	Yes	Block (2008)
535	Tamsulosin	0	0	0	0	1	1	Yes	Bonnabry (2001)
536	Telithromycin	0	0	0	0	0	1	No	Block (2008)
537	Temazepam	0	0	0	0	0	1	No	Block (2008)
538	Teniposide	0	1	0	0	0	0	No	Block (2008)
539	Tenoxicam	0	0	0	1	1	0	No	Block (2008)
540	Terfenadine	0	0	0	0	0	1	No	Block (2008)
541	Testosterone	0	0	0	0	0	1	No	Block (2008)
542	Theophylline	1	1	0	0	0	0	Yes	Block (2008)
543	Thioridazine	0	1	0	0	1	0	Yes	Block (2008)
544	Timolol	0	0	0	0	1	0	No	Block (2008)
545	Tinidazole	0	0	0	0	0	1	No	Block (2008)
546	Tipranavir	0	0	0	0	0	1	No	Block (2008)
547	Tizanidine	1	0	0	0	0	0	No	Block (2008)
548	Tolbutamide	0	1	0	1	0	0	Yes	Block (2008)
549	Tolterodine	0	0	0	0	1	1	Yes	Block (2008)

(Continued)

No.	Name	CYP1A2	CYP2C19	CYP2C8	CYP2C9	CYP2D6	CYP2E1	CYP3A4	Multilabel	Ref.
550	Torasemide	0	0	0	1	0	0	0	No	Block (2008)
551	Toremifene	0	0	0	0	0	0	1	No	Block (2008)
552	Tramadol	0	0	0	0	1	0	1	Yes	Block (2008)
553	Trazodone	0	0	0	0	1	0	1	Yes	Block (2008)
554	Triazolam	0	0	0	0	0	0	1	No	Block (2008)
555	Trimethoprim	0	0	0	1	0	0	0	No	Manga (2005)
556	Trimetrexate	0	0	0	0	0	0	1	No	Block (2008)
557	Trimipramine	0	0	0	0	1	0	0	No	Bonnabry (2001)
558	Tropisetron	0	0	0	0	1	0	1	Yes	Bonnabry (2001)
559	Valdecoxib	0	0	0	1	0	0	1	Yes	Block (2008)
560	Valsartan	0	0	0	1	0	0	0	No	Block (2008)
561	Vardenafil	0	0	0	1	0	0	1	Yes	Block (2008)
562	Venlafaxine	0	0	0	0	1	0	0	No	Block (2008)
563	Verapamil	1	0	1	0	0	0	1	Yes	Block (2008)
564	Vinblastine	0	0	0	0	0	0	1	No	Block (2008)
565	Vincristine	0	0	0	0	0	0	1	No	Block (2008)
566	Vindesine	0	0	0	0	0	0	1	No	Bonnabry (2001)
567	Vinorelbine	0	0	0	0	0	0	1	No	Block (2008)
568	Voriconazole	0	1	0	1	0	0	1	Yes	Block (2008)
569	Warfarin-(R)	1	1	1	0	0	0	1	Yes	Block (2008)
570	Warfarin-(S)	0	0	1	1	0	0	0	No	Block (2008)
571	Zafirlukast	0	0	0	1	0	0	0	No	Block (2008)
572	Zaleplon	0	0	0	0	0	0	1	No	Block (2008)
573	Zidovudine	0	0	0	0	0	0	1	No	Manga (2005)
574	Zileuton	1	0	0	1	0	0	1	Yes	Block (2008)
575	Ziprasidone	0	0	0	0	0	0	1	No	Block (2008)
576	Zolmitriptan	1	0	0	0	0	0	0	No	Block (2008)

577	Zolpidem	0	0	0	0	0	1	No	Block (2008)
578	Zonisamide	0	0	0	0	0	1	No	Block (2008)
579	Zopiclone	0	1	0	0	0	0	No	Block (2008)
580	Zuclopenthixol	0	0	0	1	0	0	No	Drug Interaction

Source: Michielan et al. (2009). *Journal of Chemical Information and Modeling* 49: 12. With permission from the American Chemical Society.

Appendix III: A 143-Member VOC Data Set and Corresponding Observed and Predicted Values of Air-to-Blood Partition Coefficients

Training Set	Name	K_{Exp}	K_{ANN}	K_{MLR}	$K_{ANN}-K_{Exp}$
1	1,1,1,2-Tetrachloroethane	1.55	1.45	0.97	−0.10
2	1,1,1-Trichloroethane	0.63	0.69	0.93	0.06
3	1,1,2,2-Tetrachloroethane	2.13	2.21	0.97	0.08
4	1,1,2-Trichloroethane	1.67	1.58	1.04	−0.09
5	1,2-Dichloroethane	1.39	1.32	1.05	−0.07
6	1-Nitropropane	2.31	2.41	2.97	0.10
7	1-Butanol	3.08	3.07	2.67	−0.01
8	1-Chloropropane	0.59	0.52	0.89	−0.07
9	1-Pentanol	2.83	3.03	2.66	0.20
10	2,2,4-Trimethylpentane	0.23	0.24	0.92	0.01
11	2-Chloropropane	0.32	0.31	0.92	−0.01
12	2-Methyl-1-propanol	2.92	2.98	2.71	0.06
13	2-Nitropropane	2.23	2.08	2.97	−0.15
14	2-Propanol	3.02	2.86	2.81	−0.16
15	4-Chlorobenzotrifluoride	1.43	1.22	2.42	−0.21
16	4-Methyl-2-pentanone	1.96	2.03	1.76	0.07
17	Propanone	2.36	2.55	1.35	0.19
18	Bromochloromethane	1.21	1.23	0.47	0.02
19	Bromodichloromethane	1.49	1.45	0.83	−0.04
20	Butyl acetate	1.94	1.71	2.34	−0.23
21	Butan-2-one	2.24	2.21	1.58	−0.03
22	Chlorodibromomethane	1.88	1.76	1.00	−0.12
23	Trichloromethane	1.15	1.13	1.31	−0.02
24	*Cis*-1,2,-dichloroethane	1.16	1.21	1.07	0.05
25	Cyclohexane	0.17	0.19	0.37	0.02
26	Decan	1.47	1.64	1.68	0.17
27	Dichloromethane	1.12	0.98	0.23	−0.14
28	Diethyl ether	1.11	1.11	1.96	0.00
29	Ethanol	3.27	3.33	2.97	0.06

(Continued)

Training Set	Name	K_{Exp}	K_{ANN}	K_{MLR}	$K_{ANN}-K_{Exp}$
30	Ethyl acetate	1.90	1.78	2.07	−0.12
31	Ethylene oxide	1.80	1.71	1.75	−0.09
32	Hexane	0.21	0.23	0.59	0.02
33	Isopentyl acetate	1.79	1.85	2.54	0.06
34	Isopropyl acetate	1.55	1.63	2.13	0.08
35	2-Bromopropane	0.64	0.61	1.05	−0.03
36	Methanol	3.41	3.44	3.33	0.03
37	Methyl *tert*-butyl ether	1.18	1.05	1.72	−0.13
38	Chloromethane	0.31	0.29	0.07	−0.02
39	Nonane	1.17	1.05	1.10	−0.12
40	*o*-Xylene	1.42	1.47	1.61	0.05
41	Pentyl acetate	1.98	2.03	2.45	0.05
42	Propyl acetate	1.88	1.75	2.07	−0.13
43	1-Bromopropane	0.97	0.93	1.00	−0.04
44	*p*-Xylene	1.61	1.81	1.89	0.20
45	Styrene	1.67	1.80	2.27	0.13
46	*Tert*-amyl methyl ether	1.22	1.09	1.82	−0.13
47	Tetrachloroethene	1.19	1.35	1.36	0.16
48	Toluene	1.14	1.18	1.89	0.04
49	Trans-1,2-dichloroethene	0.88	0.92	1.13	0.04
50	Tribromomethane	2.15	2.13	1.38	−0.02
51	Vinyl bromide	0.49	0.51	0.91	0.02
52	Vinyl chloride	0.17	0.18	0.65	0.01
53	1,2,3-Trichloropropane	2.01	1.97	1.55	−0.04
54	1,2,3-Trimethylbenzene	1.82	1.69	1.54	−0.13
55	1,2-Dichlorobenzene	2.63	2.48	1.69	−0.15
56	1,3,5-Trimethylbenzene	1.64	1.64	1.67	0.00
57	1,3-Dichlorobenzene	2.30	2.47	1.64	0.17
58	1-Chlorobutane	0.63	0.59	0.57	−0.04
59	1-Chloropentane	0.87	0.80	1.48	−0.07
60	1-Methoxy-2-propanol	4.09	4.05	3.54	−0.04
61	2-Ethoxyethanol	4.34	4.34	3.19	0.00
62	2-Hexanone	2.10	2.12	1.78	0.02
63	2-Isopropoxyethanol	4.16	4.31	3.16	0.15
64	2-Methoxyethanol	4.52	4.35	3.17	−0.17
65	3-Methylhexane	0.11	0.13	0.77	0.02
66	3-Pentanone	2.21	2.19	1.57	−0.02
67	Allylbenzene	1.71	1.61	2.29	−0.10
68	1,1-Difluroethane	0.87	0.82	0.18	−0.05
69	Fluroxene	0.42	0.42	0.63	0.00
70	Halopropane	0.15	0.14	0.70	−0.01
71	Cyclohexanone	3.33	3.14	1.62	0.15
72	Dimethyl ether	1.16	1.31	1.62	0.15
73	Divinyl ether	0.41	0.44	1.32	0.03
74	Ethyl formate	1.65	1.72	2.37	0.07

Training Set	Name	K_{Exp}	K_{ANN}	K_{MLR}	$K_{ANN-Exp}$
75	Ethyl *tert*-pentyl ether	1.25	1.30	1.92	0.05
76	Iodoethane	0.83	0.89	1.00	0.06
77	Isophorone	3.37	3.19	2.14	−0.18
78	3-Methylheptan-2-one	2.23	2.43	1.67	0.20
79	1,1-Dichloro-1-fluoroethane	0.32	0.34	0.65	0.02
80	1,1-Dichloroethylene	0.70	0.67	0.58	−0.03
81	1,2,4-Trifluorobenzene	0.76	0.73	1.36	−0.03
82	1,2-Difluorobenzene	0.96	0.93	1.24	−0.03
83	1,2-Dimethylcyclohexane	0.91	0.87	0.66	−0.04
84	1,2-Epoxy-3-butene	1.97	2.14	2.39	0.17
85	1,3,5-Trifluorobenzene	0.49	0.45	1.25	−0.04
86	1-Decene	1.21	1.17	1.62	−0.04
87	1-Hexanol	3.21	3.13	2.76	−0.08
88	1-Octene	1.07	1.11	1.29	0.04
89	1,1,1-Trifluoro-2,2-dichloroethane	0.61	0.64	0.49	0.03
90	2,3,4-Trimethylpentane	0.57	0.52	0.86	−0.05
91	2-Methylnonane	0.76	0.71	0.31	−0.05
92	2-Methyloctance	0.52	0.55	0.25	0.03
93	Bromobenzene	1.72	1.59	2.02	−0.13
94	Cyanoethylene oxide	3.22	3.13	2.50	−0.09
95	Cycloheptane	0.72	0.69	0.43	−0.03
96	Cyclopentane	0.24	0.26	0.39	0.02
97	Dibromomethane	1.87	1.86	0.70	−0.01
98	Difluoromethane	0.20	0.22	0.73	0.02
99	Fluorobenzene	1.06	1.07	1.17	0.01
100	Fluorochloromethane	0.71	0.75	0.89	0.04
101	Furane	0.82	0.78	1.34	−0.04
102	Hexafluorobenzene	0.39	0.42	0.75	0.03
103	2,3,4,5,6-Pentafluorotoluene	0.73	0.71	1.45	−0.02
104	Pentafluorobenzene	0.51	0.48	0.66	−0.03
105	4-Methylstyrene	2.37	2.22	2.42	−0.15
106	*Tert*-butylbenzene	1.24	1.28	2.01	0.04
107	*Tert*-butylcyclohexane	1.16	1.02	1.06	−0.14
	Test Set				
108	1,2-Dichloropane	1.14	1.18	1.35	0.04
109	2-Pentanone	2.14	2.38	1.69	0.24
110	Isobutyl acetate	1.69	1.80	2.30	0.11
111	Octane	0.68	0.72	0.95	0.04
112	2-Buthoxyethanol	3.90	4.02	3.38	0.12

(Continued)

Training Set	Name	K_{Exp}	K_{ANN}	K_{MLR}	$K_{ANN}-K_{Exp}$
113	1,2,4-Trimethylcyclohexane	0.87	0.81	1.27	−0.06
114	Chlorobenzene	1.63	1.85	1.77	0.22
115	Benzene	1.05	1.12	1.15	0.07
116	Chloroethane	0.49	0.50	0.69	0.01
117	1,4-Difluorobenzene	0.87	0.82	0.81	−0.05
118	3-Methyl-1-butanol	2.75	2.53	2.15	−0.22
119	Methyl acetate	1.98	2.06	2.45	0.08
120	Trichloroethene	1.14	1.21	1.06	0.07
121	Pentachloroethane	2.02	2.10	2.40	0.08
122	1,2,4-Trimethylbenzene	1.47	1.71	1.66	0.24
123	1-Bromo-2-chloroethane	1.60	1.44	1.34	−0.16
124	2-Methyl-2-propanol	2.68	2.48	2.75	−0.20
125	3-Methylstyrene	2.28	2.12	2.49	−0.16
	Validation Set				
126	1-Nonene	1.18	1.30	1.48	0.12
127	*Tert*-pentyl alcohol	2.59	2.44	2.61	−0.15
128	Isopropyl benzene	1.57	1.42	1.96	−0.15
129	2-Methylcyclohexanone	2.87	2.73	1.72	−0.14
130	Allyl chloride	1.24	1.12	1.28	−0.12
131	1,3-Dimethylbenzene	1.59	1.43	1.67	−0.16
132	Methoxyflurane	1.28	1.38	1.46	0.10
133	2,3-Dimethylbutane	0.78	0.72	0.61	−0.06
134	1,2-Dibromoethane	2.08	2.25	1.70	0.17
135	2-Methylheptane	0.49	0.43	0.60	−0.06
136	2-Heptanone	2.33	2.07	1.98	−0.26
137	Methylcyclohexane	0.70	0.65	0.59	−0.05
138	Ethylbenzene	1.47	1.30	1.24	−0.17
139	1,1-Dichloroethane	0.88	0.84	1.00	−0.04
140	1-Propanol	3.06	2.99	2.62	−0.07
141	Heptane	0.50	0.53	0.82	0.03
142	Propyl benzene	1.67	1.54	1.99	−0.13
143	Ethyl *tert*-butyl ether	1.07	1.18	1.07	0.11

Source: Konoz and Golmohammadi (2008). *Analytical Chimica Acta* 619: 157–164. With permission from Elsevier.

Index

Printed and bound by CPI Group (UK) Ltd, Croydon, CR0 4YY

18/10/2024

01776208-0019